Student Solutions Manual

for

Ott and Longnecker's

A First Course in Statistical Methods

Student Solutions Manual

for

Ott and Longnecker's

A First Course in Statistical Methods

Michael Longnecker
Texas A&M University

THOMSON
™
BROOKS/COLE

Australia • Canada • Mexico • Singapore • Spain • United Kingdom • United States

Printed in Canada
1   2   3   4   5   6   7   07   06   05   04   03

Printer: Webcom

ISBN: 0-534-40807-9

For more information about our products,
contact us at:
Thomson Learning Academic Resource Center
1-800-423-0563

For permission to use material from this text,
contact us by:
Phone: 1-800-730-2214
Fax: 1-800-731-2215
Web: http://www.thomsonrights.com

Cover Image: Jean Louis Batt/Getty Images

Brooks/Cole—Thomson Learning
10 Davis Drive
Belmont, CA 94002-3098
USA

Asia
Thomson Learning
5 Shenton Way #01-01
UIC Building
Singapore 068808

Australia/New Zealand
Thomson Learning
102 Dodds Street
Southbank, Victoria 3006
Australia

Canada
Nelson
1120 Birchmount Road
Toronto, Ontario M1K 5G4
Canada

Europe/Middle East/South Africa
Thomson Learning
High Holborn House
50/51 Bedford Row
London WC1R 4LR
United Kingdom

Latin America
Thomson Learning
Seneca, 53
Colonia Polanco
11560 Mexico D.F.
Mexico

Spain/Portugal
Paraninfo
Calle/Magallanes, 25
28015 Madrid, Spain

TABLE OF CONTENTS

Chapter 1: Statistics and the Scientific Method

1.1   a. The population of interest is the weight of shrimp maintained on the specific diet for a period of 6 months.

      b. The sample is the 100 shrimp selected from the pond and maintained on the specific diet for a period of 6 months.

      c. The weight gain of the shrimp over 6 months.

      d. Since the sample is only a small proportion of the whole population, it is necessary to evaluate what the mean weight may be for any other randomly selected 100 shrimp.

1.3   a. All households in the city that receive welfare support.

      b. The 400 households selected from the city welfare rolls.

      c. The number of children per household for those households in the city which receive welfare.

      d. In order to evaluate how closely the sample of 400 households matches the number of children in all households in the city receiving welfare.

1.5   a. All football helmets produced by the five companies over a given period of time.

      b. The 540 helmets selected from the output of the five companies.

      c. The amount of shock transmitted to the neck when the helmet's face mask is twisted.

      d. The neck strength of players is extremely variable for high school players. Hence, the amount of damage to the neck varies considerably from player to player for exactly the same amount of shock transmitted by the helmet.

Chapter 2: Collecting Data Using Surveys and Scientific Studies

2.1 The relative merits of the different types of sampling units depends on the availability of a sampling frame for individuals, the desired precision of the estimates from the sample to the population, and the budgetary and time constraints of the project.

2.3 The list of registered voters in the state could be used as the sampling frame for selecting the persons to be included in the sample.

2.5   a. Alumni (men only?) graduating from Yale in 1924.

     b. No. Alumni whose addresses were on file 25 years later would not necessarily be representative of their class.

     c. Alumni who <u>responded</u> to the mail survey would not necessarily be representative of those who were <u>sent</u> the questionaires. Income figures may not be reported accurately (intentionally), or may be rounded off to the nearest $5,000, say, in a self-administered questionaire.

     d. Rounding income responses would make the figure $25,111 highly unlikely. The fact that higher income respondents would be more likely to respond (bragging), and the fact that incomes are likely to be exaggerated, would tend to make the estimate too high.

2.7   a.
- Factors: Location in Orchard, Location on Tree, Time of Year
- Factor Levels: Location in Orchard: 8 Sections
  Time of Year: Oct., Nov., Dec., Jan., Feb., March, April, May
  Location on Tree: Top, Middle, Bottom
- Blocks: None
- Experimental Units: Locaton on Tree during one of the 8 months
- Measurement Units: Oranges
- Replications: For each section, time of year, location on tree, there is one experimental unit. Hence 1 rep.
- Treatments: 192 combinations of 8-Sections, 8 Months, 3 Locations on Tree: (S_i, M_j, L_k), for $i = 1, \cdots, 8; j = 1, \ldots, 8; k = 1, \ldots, 3$

     c.
- Factors: Type of Treatment
- Factor Levels: T_1, T_2
- Blocks: Hospitals, Wards
- Experimental Units: Patients
- Measurement Units: Patients
- Replications: 2 Patients per Treatment in each of the Ward/Hospital combinations
- Treatments: T_1, T_2

2.9   a. Randomized block design with blocking variable(5 Farms) and 48 treatments in a 3x4x4 factorial structure.

   b. Completely randomized design with 10 treatments (Software Programs) and 3 reps of each treatment.

   c. Latin square design with blocking variables (Position in Kiln, Day) each having 8 levels. The treatment structure is a 2x4 factorial structure (Type of Glaze, Thickness).

Chapter 3: Summarizing Data

3.1   a. Pie Chart should be plotted.

     b. Bar Graph should be plotted.

3.3 Pie chart should be plotted.

3.5 Two separate bar graphs could be plotted, one with Lap Belt Only and the other with Lap and Shoulder Belt. A single bar graph with the Lap Belt Only value plotted next to the Lap and Shoulder for each value of Percentage of Use is probably the most effective plot. This plot would clearly demonstrate that the increase in number of lives saved by using a shoulder belt increased considerably as the percentage use increased.

3.7 The plot has a bimodal shape. This would be an indication that there are two separate populations. However, the evidence is not very convincing because the individual plots were similar in shape with the exception that the New Therapy had a few times that were somewhat larger than the survival times obtained under the Standard Therapy.

3.9   a. There appears to have been a dramatic drop in verbal scores for both sexes from 1970 to 1980, with a greater drop in female scores. There appears to have been a slight drop in math scores from 1970 to 1980, then a slow steady rise from 1980 to 1996.

     b. Yes. For the years 1967-1996, the difference between males and females has remained relatively constant.

     c. The female verbal scores had a larger decrease for the years 1970-1980 than the male scores.

3.11 Stem-and-Leaf Plot should be plotted.

3.13 The plots show an upward trend from year 1 to year 5. There is a strong seasonal (cyclic) effect; the number of units sold increases dramatically in the late summer and fall months.

3.15 Mean = 283/16 = 17.6875 = 17.7,   Median = (14+15)/2 = 14.5,   Mode = 18

The median and mode are unalterered but the mean is inflated by the two large values.

3.17   a. Mean = 8.04, Median = 1.54

     b. Terrestrial: Mean = 15.01, Median = 6.03

       Aquatic: Mean = 0.38, Median = .375

     c. The mean is more sensitive to extreme values than is the median.

     d. Terrestrial: Median, because the two large values(76.50 and 41.70) results in a mean which is larger than 82% of the values in the data set.

       Aquatic: Mean or median since the data set is relatively symmetric.

3.19   a. If we use all 14 failure times, we obtain Mean > 173.7 and Median = 154. In fact, we know the mean is greater than 173.7 since the failure times for two of engines are greater than the reported times of 300 hours.

b. The median would be unchanged if we replaced the failure times of 300 with the true failure times for the two engines that did not fail. However, the mean would be increased.

3.21 Mean $= 1.7707$, Median $= 1.7083$, Mode $= 1.273$

The average of the three net group means and the mean of the complete set of measurements are the same. This will be true whenever the group have the same number of measurements, but is not true if the groups have different sample sizes. However, the average of the group modes and medians are different from the overall median and mode.

3.23 a. s $= 7.95$

b. Because the magnitude of the racers' ages is larger than that of their experience.

3.25 a. Luxury: $\bar{y} = 145.0, s = 27.6$
Budget: $\bar{y} = 46.1, s = 5.13$

b. Luxury hotels vary in quality, location, and price, whereas budget hotels are more competitive for the low-end market so prices tend to be similar.

3.27 a. Stem-and-Leaf Plot is given here:

Stem	Leaf
2	5
2	677
2	9
3	00111
3	223333
3	5
3	67
3	8

b. Min=250, $Q_1 = (295 + 301)/2 = 298, Q_2 = (315 + 320)/2 = 317.5$,
$Q_3 = (334 + 334)/2 = 334$, Max=386

A boxplot is given here.

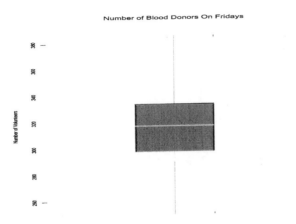

Number of Blood Donors On Fridays

There are no outliers because $Q_1 - (1.5)IQR = 244 < 250$ and $Q_3 + (1.5)IQR = 388 > 386$. The distribution is approximately symmetric with no outliers.

3.29　a. Stacked Bar Graph is given here:

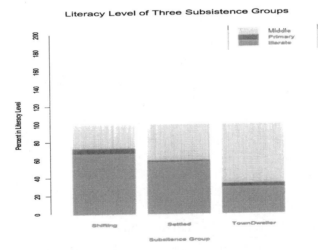

Literacy Level of Three Subsistence Groups

b. Illiterate: 46%,　Primary Schooling: 4%,　At Least Middle School: 50%
Shifting Cultivators: 27%,　Settled Agriculturists: 21%,　Town Dwellers: 51%
There is a marked difference in the distribution in the three literacy levels for the three subsistence groups. Town dwellers and shifting cultivators have the reverse trends in the three categories, whereas settled agriculturists fall into essentially two classes.

3.31 A scatterplot of M3 versus M2 is given here.

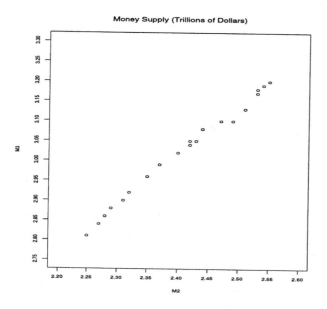

a. Yes, it would since we want to determine the relative changes in the two over the 20 month period of time.

b. See scatterplot. The two measures follow an approximately increasing linear relationship.

3.33　a. Mean = 57.5,　　　　　　　　Median = 34.0

b. Median since the data has a few very large values which results in the mean being larger than all but a few of the data values.

c. Range = 273,　　　　　　　$s = 70.2$

d. Using the approximation, $s \approx range/4 = 273/4 = 68.3$. The approximation is fairly accurate.

e. $\bar{y} \pm s \Rightarrow (-12.7, 127.7)$; yields 82%

$\bar{y} \pm 2s \Rightarrow (-82.9, 197.0)$; yields 94%

$\bar{y} \pm 3s \Rightarrow (-153.1, 268.1)$; yields 97%

These percentages do not match the Empirical Rule very well: 68%, 95%, and 99.7%

f. The Empirical Rule applies to data sets with roughly a "mound-shaped" histogram. The distribution of this data set is highly skewed right.

3.35 A scatterplot of is given here.

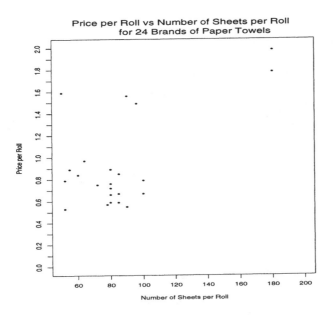

Price per Roll vs Number of Sheets per Roll
for 24 Brands of Paper Towels

Price per Roll

Number of Sheets per Roll

   a. The points do not fall close to a straight line.

   b. No, as the number of sheets increases from 50 to 100, there is just a scatter of points,
      no real pattern. The price per roll jumps dramatically for the two brands having the
      largest number of sheets.

   c. Paper towel sheets vary in thickness and size which will affect the price.

3.37   a. Construct a relative frequency histogram

        b. Highly skewed to the right

        c. $1424 \pm 3488 \Rightarrow$ (-2063, 4912) contains 37/41 = 90.2%
           $1424 \pm (2)3488 \Rightarrow$ (-5551, 8400) contains 38/41 = 92.7%
           $1424 \pm (3)3488 \Rightarrow$ (-9039, 11888) contains 39/41 = 95.1%
           These values do not match the percentages from the Empirical Rule: 68%, 95%, and
           99.7%.

        d. $1.48 \pm 1.54) \Rightarrow$ (-0.06, 3.02) contains 31/41 = 75.6%
           $1.48 \pm (2)1.54 \Rightarrow$ (-1.60, 4.56) contains 40/41 = 97.6%
           $1.48 \pm (3)1.54 \Rightarrow$ (-3.14, 6.10) contains 41/41 = 100%
           These values closely match the percentages from the Empirical Rule: 68%, 95%, and
           99.7%.

3.39   a. There has been very little change from 1985 to 1996.

        b. Yes; For both 1985 and 1996, DC had extremely low ownership. New York and Hawaii
           had semi-low ownership.

9

c. No

d. The cost of homes is very high.

3.41   a. Plot relative frequency histogram.

b. 1100

c. mean $= 25115/23 = 1091.96$,   median $= 1039$

d. Because the mean is slightly larger than the median, it is likely that the distribution is slightly skewed to the right.

3.43 The stem-and-leaf diagram is given here:

Stem	Leaf
54	7
55	
56	
57	
58	
59	
60	
61	
62	5
63	0
64	
65	6
66	4 7 7
67	79
68	8 88 88
69	1 4 7 9
70	0 1 2 333 8
71	1

Yes, the distribution is left skewed.

3.45   a. New policy $\bar{y} = 2.27$, $s = 3.26$
       Old policy: $\bar{y} = 4.60$, $s = 2.61$

b. Both the average number of sick days and the variation in number of sick days have decreased with new policy.

3.47   a. Average Price $= 76.68$

b. Range $= 202.69$

c. DJIA $= 10856.2$

d. Yes; The stocks covered on the NYSE.
   No; The companies selected are the leading companies in the different sectors of business.

3.49   a. The value 62 reflects the number of respondents in coal producing states who preferred a national energy policy that encouraged coal production. The value 32.8 is of those who favored a coal policy, 32.80% came from major coal producing states. The value 41.3 tells us that 41.3% of those from coal states favored a coal policy. And the value 7.8 tells us that 7.8% of all the responses come from residents of major coal producing states who were in favor of a national energy policy that encouraged coal production.

       b. The column percentages because they displayed the distribution of opinions within each of the three types of states.

       c. Yes. For both the coal and oil-gas states, the largest percentage of responses favored the type of energy produced in their own state.

3.51   a. A boxplot of the magnitude data is given here. The smallest outlier is approximately 2.7.

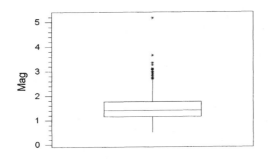

Magnitude of Earthquake

       b. There are 28 bigger earthquakes with a median magnitude of 2.725.

       c. A side-by-side boxplot and summary statistics for the magnitude of earthquakes are given here.

Descriptive Statistics: Magitude by BeforeAfter

Variable		N	Mean	Median	TrMean	StDev
Magitude	Before	230	1.4957	1.4000	1.4629	0.5545
	After	229	1.4955	1.4200	1.4691	0.5180
	Yountville	1	5.1700	5.1700	5.1700	*

Variable	BefAft	SE Mean	Minimum	Maximum	Q1	Q3
Magitude	Before	0.0366	0.5000	3.6600	1.1300	1.7800
	After	0.0342	0.5400	3.1100	1.1600	1.7350
	Yountville	*	5.1700	5.1700	*	*

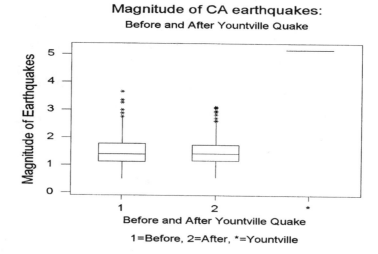

Magnitude of CA earthquakes:
Before and After Yountville Quake

Magnitude of Earthquakes

Before and After Yountville Quake

1=Before, 2=After, *=Yountville

There does not seem to be an major differences in the magnitude of earthquakes before and after the Yountville quake.

d. Summary statistics and plots are given here for earthquakes near Yountvile.

Descriptive Statistics: YLat, YLon, YMag

Variable	N	Mean	Median	TrMean	StDev	SE Mean
YLat	33	38.369	38.369	38.370	0.015	0.003
YLon	33	-122.39	-122.40	-122.39	0.01	0.00
YMag	33	1.694	1.550	1.591	0.721	0.125

Variable	Minimum	Maximum	Q1	Q3
YLat	38.309	38.399	38.363	38.376
YLon	-122.41	-122.35	-122.40	-122.38
YMag	0.990	5.170	1.295	1.780

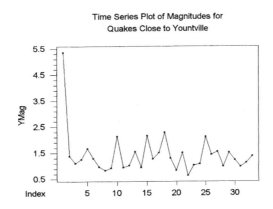

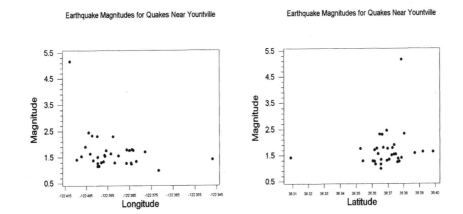

There were no recorded earthquakes close to Yountville prior to the magnitude 5.17 quake but 33 after it occurred. The 5.17 magnitude quake was nearly centered in a north-south direction with respect to the "aftershocks", but the 5.17 quake was definitely shifted to the west of the "aftershocks".

Chapter 4: Probability and Probability Distributions

4.1   a. Subjective probability

 b. Relative frequency

 c. Classical

 d. Relative frequency

 e. Subjective probability

 f. Subjective probability

 g. Classical

4.3 Using the binomial formula, the probability of guessing correctly 15 or more of the 20 questions is 0.021.

4.5   HHH, HHT, HTH, THH, TTH, THT, HTT, TTT

4.7   a. $P\left(\overline{A}\right) = 1 - \frac{3}{8} = \frac{5}{8}$

 $P\left(\overline{B}\right) = 1 - \frac{7}{8} = \frac{1}{8}$

 $P\left(\overline{C}\right) = 1 - \frac{1}{8} = \frac{7}{8}$

 b. Events A and B are not mutually exclusive because B contains A $\Rightarrow$ $A \cap B = A$, which is not the empty set.

4.9 Because $P(A|B) = \frac{3}{7} \neq \frac{3}{8} = P(A) \Rightarrow$ A and B are not independent.

 Because $P(A|C) = 0 \neq \frac{3}{8} = P(A) \Rightarrow$ A and C are not independent.

 Because $P(B|C) = 0 \neq \frac{7}{8} = P(B) \Rightarrow$ B and C are not independent.

4.11 No, since A and B are not mutually exclusive. Also, $P(A \cap B) = 0.30 \neq 0$

4.13   a. $S = \{F_1F_2, F_1F_3, F_1F_4, F_1F_5, F_2F_1, F_2F_3, F_2F_4, F_2F_5, F_3F_1, F_3F_2,$
 $F_3F_4, F_3F_5, F_4F_1, F_4F_2, F_4F_3, F_4F_5, F_5F_1, F_5F_2, F_5F_3, F_5F_4\}$

 b. Let T_1 be the event that the 1st firm chosen is stable and T_2 be the event that the 2nd firm chosen is stable.

 $P(T_1) = \frac{3}{5}$ and $P(\overline{T_1}) = \frac{2}{5}$

 $P(\text{Both Stable}) = P(T_1 \cap T_2) = P(T_2|T_1)P(T_1) = (\frac{2}{4})(\frac{3}{5}) = \frac{6}{20} = 0.30$

 Alternatively, if we designated F_1 and F_2 as the Shakey firms, then we could go to the list of 20 possible outcomes and identify 6 pairs containing just Stable firms: $F_3F_4, F_3F_5, F_4F_3, F_4F_5, F_5F_3, F_5F_4$. Thus, the probability that both firms are Stable is $\frac{6}{20}$.

c. $P(\text{One of two firms is Shakey})$
$$= P(\text{1st chosen is Shakey and 2nd chosen is Stable})$$
$$+P(\text{1st chosen is Stable and 2nd chosen is Shakey})$$
$$= P(\overline{T_1} \cap T_2) + P(T_1 \cap \overline{T_2}) = P(T_2|\overline{T_1})P(\overline{T_1}) + P(\overline{T_2}|T_1)P(T_1)$$
$$= (\tfrac{3}{4})(\tfrac{2}{5}) + (\tfrac{2}{4})(\tfrac{3}{5}) = \tfrac{12}{20} = 0.60$$

Alternatively, in the list of 20 outcomes there are 12 pairs which consist of exactly 1 Shakey firm (F_1 or F_2) and exactly 1 Stable firm (F_3 or F_4 or F_5). Thus, the probability of exactly 1 Shakey firm is $\tfrac{12}{20}$.

d. $P(\text{Both Shakey}) = P(\overline{T_1} \cap \overline{T_2}) = P(\overline{T_2}|\overline{T_1})P(\overline{T_1}) = (\tfrac{1}{4})(\tfrac{2}{5}) = \tfrac{2}{20} = 0.10$

Alternatively, in the list of 20 outcomes there are 2 pairs in which both firms are Shakey (F_1 or F_2). Thus, the probability that both firms are Shakey is $\tfrac{2}{20}$.

4.15   a. $P(A) = P(none \cap high) + P(little \cap high) + P(some \cap high) + P(extensive \cap high)$
$$= 0.10 + 0.15 + 0.16 + 0.22 = 0.63$$
$$P(B) = P(low \cap extensive) + P(medium \cap extensive) + P(high \cap extensive)$$
$$= 0.04 + 0.10 + 0.22 = 0.36$$
$$P(C) = P(low \cap none) + P(low \cap little) + P(medium \cap none) + P(medium \cap little)$$
$$= 0.01 + 0.02 + 0.05 + 0.06 = 0.14$$

b. $P(A|B) = P(A \cap B)/P(B) = 0.22/0.36 = 0.611$
$P(A|B) = P(A \cap B)/P(B) = 0.41/(1 - .36) = 0.64$
$P(B|C) = P(B \cap C)/P(C) = 0.14/0.14 = 1$

c. $P(A \cup B) = P(A) + P(B) - P(A \cap B) = 0.63 + 0.36 - 0.22 = 0.77$
$P(A \cap C) = 0; \quad P(B \cap C) = 0$

4.17 Let A = event customer pays 1st month's bill in full and B = event customer pays 2nd month's bill in full. We are given that

$$P(A) = 0.70, P(B|A) = 0.95, P(\overline{B}|\overline{A}) = 0.90, P(B|\overline{A}) = 1 - P(\overline{B}|\overline{A}) = 1 - 0.90 = 0.10$$

a. $P(A \cap B) = P(B|A)P(A) = (0.95)(0.70) = 0.665$

b. $P(\overline{B} \cap \overline{A}) = P(\overline{B}|\overline{A})P(\overline{A}) = (0.90)(1 - 0.70) = 0.270$

c. $P(\text{pay exactly one month in full})$
$$= 1 - P(\text{pays neither month or pays both months})$$
$$= 1 - P(\overline{A} \cap \overline{B}) - P(A \cap B) = 1 - 0.665 - 0.27 = 0.065$$

4.19 $P(D|R_2) = \dfrac{P(R_2|D)P(D)}{P(R_2|D)P(D)+P(R_2|\overline{D})P(\overline{D})} = \dfrac{(0.40)(0.01)}{(0.40)(0.01)+(0.40)(0.99)} = 0.01$

The two probabilities are equal since the proportion of fair risk applicants is the same for both the defaulted and nondefaulted loans.

4.21   a. Sensitivity: $P(DA|C) = 50/53 = 0.943$
Specificity: $P(DNA|RO) = 44/47 = 0.936$

b. $P(C|DA) = \dfrac{(0.943)(0.00108)}{(0.943)(0.00108)+(0.021)(1-0.00108)} = 0.046$

c. $P(RO|DA) = 1 - P(C|DA) = 1 - 0.046 = .954$

d. $P(RO|DNA) = \frac{(0.936)(1-0.00108)}{(0.936)(1-0.00108)+(0.019)(0.00108)} = 0.99998$

4.23   a. a. $P(y \leq 4) = P(0) + P(1) + P(2) + P(3) + P(4)$
$$= 0.0168 + 0.0896 + 0.2090 + 0.2787 + 0.2322 = 0.8263$$

b. $P(y > 4) = 1 - P(y \leq 4) = 1 - 0.8263 = 0.1737$

c. $P(y \leq 7) = 1 - P(8) = 1 - \binom{8}{8}(.4)^8(.6)^{8-8} = 1 - 0.0007 = 0.9993$

d. $P(y > 6) = P(y = 7) + P(y = 8) = \binom{8}{7}(.4)^7(.6)^{8-7} + \binom{8}{8}(.4)^8(.6)^{8-8} = 0.0085$

4.25   a. $P(y \geq 3) = 1 - P(y < 3) = 1 - (0.06 + 0.14 + 0.16) = 0.64$

b. $P(2 \leq y \leq 6) = P(y = 2) + P(y = 3) + P(y = 4) + P(y = 5) + P(y = 6)$
$$= 0.16 + 0.14 + 0.12 + 0.10 + 0.08 = 0.60$$

c. $P(y > 8) = P(y = 9) + P(y = 10) = 0.04 + 0.03 = 0.07$

4.27 No, people may not answer the question.

4.29 Binomial experiment with n=10 and $\pi = 0.30$.

a. $P(y = 0) = \binom{10}{0}(.3)^0(.7)^{10} = 0.0282$
$P(y = 6) = \binom{10}{6}(.3)^6(.7)^4 = 0.0368$
$P(y \geq 6) = 1 - P(y < 6)$
$$= 1 - (P(0) + P(1) + P(2) + P(3) + P(4) + P(5)) = 1 - 0.9527 = 0.0473$$
$P(y = 10) = \binom{10}{10}(.3)^{10}(.7)^0 = 0.000006$

b. With $n = 1000$ and $\pi = .3$ $P(y \leq 100) = \sum_{i=0}^{100} \binom{1000}{i}(.3)^i(.7)^{1000-i}$.
This would be a length calculation. In Section 4.13, we will provide an approximation which will greatly reduce the amount of calculations. However, we can also consider that
$\mu = n\pi = (1000)(.3) = 300, \sigma = \sqrt{n\pi(1-\pi)} = \sqrt{1000(.3)(.7)} = 14.49$
$\mu \pm 3\sigma = 300 \pm (3)(14.49) = (256.53, 343.47)$.
Thus, the chance of observing the event $y \leq 100$ is very small.

4.31 No. The trials are not identical.

4.33   a. 0.9032 - 0.5000 = 0.4032

b. 0.5000 - 0.0287 = 0.4713

4.35   a. 0.9115 - 0.4168 = 0.4947

b. 0.8849 - 0.6443 = 0.2406

4.37 $z_o = 1.96$

4.39 $z_o = 1.645$

4.41   a. $P(500 < y < 696) = P(\frac{500-500}{100} < z < \frac{696-500}{100})$
$$= P(0 < z < 1.96) = 0.9750 - 0.5000 = 0.475$$

b. $P(y > 696) = P(z > \frac{696-500}{100}) = P(z > 1.96) = 1 - 0.9750 = 0.025$

c. $P(304 < y < 696) = P(\frac{304-500}{100} < z < \frac{696-500}{100})$
$$= P(-1.96 < z < 1.96) = 0.9750 - 0.0250 = 0.95$$

d. $P(500 - k < y < 500 + k) = P(\frac{500-k-500}{100} < z < \frac{500+k-500}{100})$
$$= P(-.01k < z < .01k) = 0.60.$$
From Table 1 we find,
$$P(-.845 < z < .845) = 0.60. \text{ Thus, } .01k = .845 \Rightarrow k = 84.5$$

4.43 a. $z = 2.33$

b. $z = -1.28$

4.45 a. $\mu = 39; \sigma = 6; P(y > 50) = P(z > \frac{50-39}{6}) = P(z > 1.83) = 1 - 0.9664 = 0.0336$

b. Since 55 is $\frac{55-39}{6} = 2.67$ std. dev. above $\mu = 39$, thus $P(y > 55) = P(z > 2.67)$
$= 0.0038$. We would then conclude that the voucher has been lost.

4.47 $\mu = 150; \quad \sigma = 35$

a. $P(y > 200) = P(z > \frac{200-150}{35}) = P(z > 1.43) = 0.0764$

b. $P(y > 220) = P(z > \frac{220-150}{35}) = P(z > 2) = 0.0228$

c. $P(y < 120) = P(z < \frac{120-150}{35}) = P(z < -0.86) = 0.1949$

d. $P(100 < y < 200) = P(\frac{100-150}{35} < z < \frac{200-150}{35}) = P(-1.43 < z < 1.43) = 0.8472$

4.49 No. The sample would be biased toward homes for which the homeowner is at home much of the time. For example, the sample would tend to include more people who work at home and retired persons.

4.51 Start at column 2 line 1 we obtain 150, 729, 611, 584, 255, 465, 143, 127, 323, 225, 483, 368, 213, 270, 062, 399, 695, 540, 330, 110, 069, 409, 539, 015, 564. These would be the women selected for the study.

4.53 The sampling distribution would have a mean of 60 and a standard deviation of $\frac{5}{\sqrt{16}} = 1.25$. If the population distribution is somewhat mound-shaped then the sampling distribution of $\bar{y}$ should be approximately mound-shaped. In this situation, we would expect approximately 95% of the possible values of $\bar{y}$ is lie in $60 \pm (2)(1.25) = (57.5, 62.5)$.

4.55 $\mu = 930; \quad \sigma = 130$

a. $P(800 < y < 1100) = P(\frac{800-930}{130} < z < \frac{1100-930}{130}) = P(-1 < z < 1.31)$
$$= 0.9049 - 0.1587 = 0.7462$$

b. $P(y < 800) = P(z < \frac{800-930}{130}) = 0.1587$

c. $P(y > 1200) = P(z > \frac{1200-930}{130}) = P(z > 2.08) = 1 - 0.9811 = 0.0189$

4.57 $\mu = 2.1; \quad \sigma = 0.3$

a. $P(y > 2.7) = P(z > \frac{2.7-2.1}{0.3}) = P(z > 2) = 0.0228$

b. $P(z > 0.6745) = 0.25 \Rightarrow y_{.75} = 2.1 + (0.6745)(0.3) = 2.30$

c. Let μ_N be the new value of the mean. We need $P(y > 2.7) \le 0.05$.
From Table 1, $0.05 = P(z > 1.645)$ and $0.05 = P(y > 2.7) = P(\frac{y-\mu_N}{.3} > \frac{2.7-\mu_N}{0.3}) \Rightarrow$
$\frac{2.7-\mu_N}{0.3} = 1.645 \Rightarrow \mu_N = 2.7 - (0.3)(1.645) = 2.2065$

4.59 Individual baggage weight has $\mu = 95$; $\sigma = 35$;   Total weight has mean $n\mu = (200)(95) = 19,000$; and standard deviation $\sqrt{n}\sigma = \sqrt{200}(35) = 494.97$. Therefore, $P(y > 20,000) = P(z > \frac{20,000-19,000}{494.97}) = P(z > 2.02) = 0.0217$

4.61   a. $\mu = n\pi = (10000)(0.001) = 10$

        b. $\sigma = \sqrt{10000)(0.001)(0.999}} = 3.16$

            $P(y < 5) = P(y \leq 4) \approx P(z \leq \frac{4.5-10}{3.16}) = P(z < -1.74) = 0.0409$

        c. $P(y < 2) = P(y \leq 1) = P(z \leq \frac{1.5-10}{3.16}) = P(z < -2.69) = 0.0036$

4.63 $\sigma_{\bar{y}} = \frac{10.2}{\sqrt{15}} = 2.63$

4.65 No, there is strong evidence that the new fabric has a greater mean breaking strength.

4.67 Could use random sampling by first identifying all returns with income greater than \$15,000. Next, create a computer file with all such returns listed in alphabetical order by last name of client. Randomly select 1% of these returns using a systematic random sampling technique.

4.69 $n = 400$ $\pi = 0.2$

        a. $\mu = n\pi = 400(.20) = 80; \sigma = \sqrt{400(.2)(.8)} = 8$

            $P(y \leq 25) \approx P(z \leq \frac{25-80}{8}) = P(z \leq -6.875) \approx 0.$

        b. The ad is not successful. With $\pi = .20$, we expect 80 positive responses out of 400 but we observed only 25. The probability of getting so few positive responses is virtually 0 if $\pi = .20$. We therefore conclude that π is much less than 0.20.

4.71 a.,b. The mean and standard deviation of the sampling distribution of $\bar{y}$ are given when the population distribution has values $\mu = 100, \sigma = 15$:

Sample Size	Mean	Standard Deviation
5	100	6.708
20	100	3.354
80	100	1.677

        c. As the sample size increases, the sampling distribution of $\bar{y}$ concentrates about the true value of μ. For $n = 5$ and 20, the values of $\bar{y}$ could be a considerable distance from 100.

5.1   a. All registered voters in the state.

b. Simple random sample from a list of registered voters.

5.3 It would depend on many factors relating to the manner and locations in which the fuses are manufactured and stored.

5.5   a. The width of the interval will be decreased.

b. The width of the interval will be increased.

5.7   a. $10.4 \pm (2.58)(\frac{4.2}{\sqrt{400}}) = 10.4 \pm 0.54 = (9.86, 10.94)$

b. No, we are 99% confident that the average tire pressure is between 9.86 and 10.94 psi. underflated. Since a tire is considered to be seriously underinflated only if its tire pressure is more than 10 psi underinflated, this is a good chance that tires may not be seriously underinflated.

c. The 90% C.I. is $10.4 \pm (1.645)(\frac{4.2}{\sqrt{400}}) = 10.4 \pm 0.35 = (10.05, 10.75)$. Yes, since this interval is completely above 10 psi.

5.9 $850 \pm (1.96)(\frac{100}{\sqrt{60}}) = 850 \pm 25.3 = (824.7, 875.3)$

5.11   a. $n = \frac{(1.96)^2 (2.5)^2}{(1/2)^2} = 97$

b. $n = \frac{(2.58)^2 (2.5)^2}{(1.2/2)^2} = 116$

c. $n = \frac{(1.645)^2 (2.5)^2}{(1.2/2)^2} = 47$

5.13 $\hat{\sigma} = 13, E = 3, \alpha = .01 \Rightarrow n = \frac{(2.58)^2 (13)^2}{(3)^2} = 125$

5.15   a. $\hat{\sigma} = \frac{1500 - 200}{4} = 325 \Rightarrow n = \frac{(2.58)^2 (325)^2}{(25)^2} = 1125$

b. The 95% level of confidence implies that there will be a 1 in 20 chance, over a large number of samples, that the confidence interval will not contain the population average rent. The 99% level of confidence implies there is only a 1 in 100 chance of not containing the average. Thus, we would increase the odds of not containing the true average five-fold.

5.17 a.-c. The power values, $PWR(\mu_a)$, are given here:

		μ_a					
n	α	39	40	41	42	43	44
50	0.05	0.3511	0.8107	0.9840	0.9997	1.0000	1.0000
50	0.025	0.2428	0.7141	0.9662	0.9990	0.9999	1.0000
20	0.05	0.1987	0.4809	0.7736	0.9394	0.9906	0.9992

The power curves are plotted here:

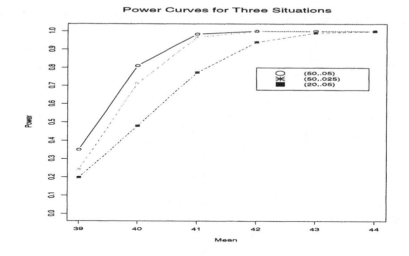

Power Curves for Three Situations

Legend: (50,.05), (50,.025), (20,.06)

5.19 $H_o : \mu \leq 2$ vs $H_a : \mu > 2, \bar{y} = 2.17, s = 1.05, n = 90$

    a. $z = \frac{2.17-2}{1.05/\sqrt{90}} = 1.54 < 1.645 = z_{0.05} \Rightarrow$

    Fail to reject H_o. The data does not support the hypothesis that the mean has been decreased from 2.

    b. $\beta(2.1) = P\left(z \leq 1.645 - \frac{|2-2.1|}{1.05/\sqrt{90}}\right) = P(z \leq 0.74) = 0.7704$

5.21 $n = \frac{(80)^2(1.645+1.96)^2}{(525-550)^2} = 133.1 \Rightarrow n = 134$

5.23 $H_o : \mu \leq 30$ vs $H_a : \mu > 30$

    $\alpha = 0.05, n = 37, \bar{y} = 37.24, s = 37.12$

    a. $z = \frac{37.24-30}{37.12/\sqrt{37}} = 1.19 < 1.645 = z_{0.05} \Rightarrow$ Fail to reject H_o.

    There is not sufficient evidence to conclude that the mean lead concentration exceeds 30 mg kg^{-1} dry weight.

    b. $\beta(50) = P(z \leq 1.645 - \frac{|30-50|}{37.12/\sqrt{37}}) = P(z \leq -1.63) = 0.0513$.

    c. No, the data values are not very close to the straight-line in the normal probability plot.

    d. No, since there is a substantial deviation from a normal distribution, the sample size should be somewhat larger to use the $z-$ test. Section 5.8 provides an alternative test statistic for handling this situation.

5.25 p-value $= 0.0359 > 0.025 = \alpha \Rightarrow$

No, there is not significant evidence that the mean is greater than 45. With $\alpha = 0.025$, the researcher is demanding greater evidence in the data to support the research hypothesis.

5.27 $H_o : \mu = 1.6$ versus $H_a : \mu \neq 1.6$,

$n = 36, \bar{y} = 2.2, s = 0.57, \alpha = 0.05$.

p-value $= 2P(z \geq \frac{|2.2-1.6|}{.57/\sqrt{36}}) = 2P(z \geq 6.32) < 0.0001 < 0.05 = \alpha \Rightarrow$

Yes, there is significant evidence that the mean time delay differs from 1.6 seconds.

5.29   a. Reject H_o if $t \leq -2.624$

   b. Reject H_o if $|t| \geq 2.819$

   c. Reject H_o if $t \geq 3.365$

5.31 $H_o : \mu \leq 80$ versus $H_a : \mu > 80, n = 20, \bar{y} = 82.05, s = 10.88$

$t = \frac{82.05-80}{10.88/\sqrt{20}} = 0.84 \Rightarrow$ Reject H_o if $t \geq 1.729$.

Fail to reject H_o and conclude data does not support the hypothesis that the mean reading comprehension is greater than 80.

The level of significance is given by p-value $= P(t \geq 0.84) \approx 0.20$.

5.33   a. Yes, the plotted points are all close to the straight line.

   b. Fairly close.

   c. Yes, since the confidence interval contains values which are greater than 35. However, the C.I. is a two-sided procedure which yields a one-sided α of 0.005 which would require greater evidence from the data to support the research hypothesis.

5.35   a. Untreated: $43.6 \pm (1.833)(5.7)/\sqrt{10} \Rightarrow (40.3, 46.9)$
   Treated: $36.1 \pm (1.833)(4.9)/\sqrt{10} \Rightarrow (33.3, 38.9)$

   We are 90% confident that the average height of untreated shrups is between 40.3 cm and 46.9 cm. We are 90% confident that the average height of treated shrups is between 33.3 cm and 38.9 cm.

   b. The two intervals do not overlap. This would indicate that the average heights of the treated and untreated shrups are significantly different.

5.37 Reject H_o if $B \geq 20$

5.39   a. The normal probability plot is given here:

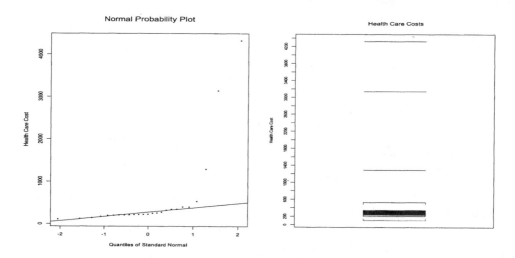

The data set does not appear to be a sample from a normal distribution, since a large proportion of the values are outliers as depicted in the box plot and several points are a considerable distance from the line in the normal probability plot. The data appears to be from an extremely right skewed distribution.

b. Because of the skewness, the median would be a better choice than the mean.

c. (208, 342) We are 95% confident that the median amount spent on healthcare by the population of hourly workers is between $208 and $342 per year.

d. Reject $H_o : M \leq 400$ if $B \geq 25 - 7 = 18$.

   We obtain B = 4; Since 4 < 18, do not reject $H_o : M \leq 400$. The data fails to demonstrate that the median amount spend on health care is greater than $400.

5.41   a. Reject $H_o : M \leq 0.25$ in favor of $H_a : M > 0.25$ at level $\alpha = 0.01$ if $B \geq 25 - 6 = 19$. From the differences, $y_i - 0.25$, we obtain $B = 18$ positive values. Thus, we fail to reject H_o and conclude that the data does not support a median increase in reaction time of at least 0.25 seconds.

   b. Weight of driver, experience(age) of driver, amount of sleep in previous 24 hours, etc.

5.43   a. $\bar{y} = 1.466$

   b. 95% C.I.: $1.466 \pm (2.145)(.3765)/\sqrt{15} \Rightarrow (1.26, 1.67)$

      We are 95% confident that the average mercury content after the accident is between 1.28 and $1.66 mg/m^3$

   c. $H_a : \mu > 1.20$ Reject H_o if $t \geq 1.761$.
      $t = \frac{1.466 - 1.2}{.3765/\sqrt{15}} = 2.74 \Rightarrow$

      There is sufficient evidence that the mean mercury concentration has increased.

   d. Using Table 4, we obtain the following with $d = \frac{|\mu_a - 1.2|}{.32}$

μ_a	d	$PWR(\mu_a)$
1.28	0.250	0.235
1.32	0.375	0.396
1.36	0.500	0.578
1.40	0.625	0.744

5.45 $H_0 : \mu \geq 300$ versus $H_a : \mu < 300$,

$n = 20, \bar{y} = 160, s = 90, \alpha = 0.05$

p-value $= P(t \leq \frac{160-300}{90/\sqrt{20}}) = P(t \leq -6.95) < 0.0001 < 0.05 = \alpha.$

Yes, there is sufficient evidence to conclude that the average is less than \$300.

5.47   a. $\bar{y} = 74.2$ 95% C.I.: $74.2 \pm (2.145)(44.2)/\sqrt{15} \Rightarrow (49.72, 98.68)$

b. $H_0 : \mu \leq 50$ versus $H_a : \mu > 50$,

$n = 15, \alpha = 0.05$

p-value $= P(t \geq \frac{74.2-50}{44.2/\sqrt{15}}) = P(t \geq 2.12) = 0.0262 < 0.05 = \alpha.$

Yes, there is sufficient evidence to conclude that the average daily output is greater than 50 tons of ore.

5.49   a. The summary statistics are given here:

Time	Mean	Std.Dev	n	95% C.I.
6 A.M.	0.128	0.0355	15	(0.108, 0.148)
2 P.M.	0.116	0.0406	15	(0.094, 0.138)
10 P.M.	0.142	0.0428	15	(0.118, 0.166)
All Day	0.129	0.0403	45	(0.117, 0.141)

b. No, the three C.I.'s have a considerable overlap.

c. $H_o : \mu \geq 0.145$ versus $H_a : \mu < 0.145$

p-value $= P(t \leq \frac{.129-.145}{.0403/\sqrt{45}}) = P(t \leq -2.66) = 0.0054$

There is significant evidence (very small p-value) that the average SO_2 level using the new scrubber is less than 0.145.

5.51 $n = \frac{(12.36)^2(1.96)^2}{(1)^2} = 586.9 \Rightarrow n = 587$

5.53 $n = 40, \bar{y} = 58, s = 10$

99% C.I. on μ : $58 \pm (2.708)(10)/\sqrt{40} \Rightarrow (53.7, 62.3)$

5.55 From the material in the Encounters with Real Data section, we obtain the following answers.

a.   1. The population from which the data was randomly selected is a population of female nurses.

2. Some possible dietary variables are total calories consumed per day; amount of salt, amount of protein, amount of fiber in the diet; amount of liquids consumed; and many other factors.

3. Percent body fat, amount of daily exercise, blood pressure, family health history, age, and many other factors.

4. Obtain a list of names of all nurses in the population from which the sample is to be selected. Assign a number to each name. Randomly select 168 numbers from 1 to N, the total number of nurses in the population, using a computer program or the random number table in the Appendix. The nurses corrsponding to each of these 168 numbers will be in the study.

5. Does the population of nurses have a higher proportion with PCF above 50% than populations of other professionals, blue collar workers, etc.? Is there greater variability in the PCF values for nurses in comparison to populations of other groups of people?

Chapter 6: Inferences Comparing Two Population Central Values

6.1   a. Reject H_o if $|t| \geq 2.064$

       b. Reject H_o if $t \geq 2.624$

       c. Reject H_o if $t \leq -1.860$

6.3 p-value $= P(t \leq -2.6722) \Rightarrow 0.005 <$ p-value < 0.01

6.5   a. $H_o : \mu_A = \mu_B$ versus $H_a : \mu_A \neq \mu_B$; p-value $= 0.065 \Rightarrow$ The data do not provide sufficient evidence to conclude there is a difference in mean oxygen content.

       b. The separate variance t'-test was used since it has df given by
$$c = \frac{(.157)^2/15}{(.157)^2/15 + (.320)^2/15} = 0.194; \ df = \frac{(15-1)(15-1)}{(1-.194)^2(15-1) + (.194)^2(15-1)} \approx 20.$$
If the pooled t-test was used the df=15+15-2=28 but the separate variance test has $df \approx 20$, as is shown on printout.

       c. The Above-town and Below-town data sets appear to be normally distributed, with the exception that the box-plot does display an outlier for the Below-town data. The Below-town data appears to be more variable than the Above-town data. The two data sets consist of independent random samples.

       d. The 95% C.I. estimate for the difference in means is (-0.013, 0.378) ppm. The observed difference (0.183 ppm) is not significant.

6.7 We want to test $H_o : \mu_{No} = \mu_{Sub}$ versus $H_a : \mu_{No} \neq \mu_{Sub}$; p-value $= 0.0049 \Rightarrow$ The data supports the contention that people that receive a daily newspaper have a greater knowledge of current events. The stem-and-leaf plots indicate that both data sets are from normally distributed populations but with different variances. The problem description indicates that the two random samples were independently selected. A 95% C.I. on the mean difference is (-15.5, -2.2) which reflects the lower values from the people who did not read a daily newspaper.

6.9 Since the sample sizes are very small, we would need to be assured that the samples were randomly selected from normally distributed populations.

6.11 If the data were analyzed using the difference in PCB content (1996 data - 1982 data) at each site, the effect of between site variability could potentially be reduced. The data should be analyzed as a *before and after* study using paired data methodology.

6.13   a. Plumber 1: $\bar{y}_1 = 88.81$, $s_1 = 7.89$

          Plumber 2: $\bar{y}_2 = 108.93$, $s_2 = 8.73$

       b. The box plots are given here:

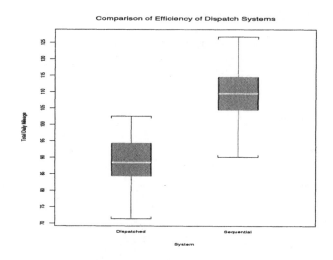

Comparison of Efficiency of Dispatch Systems

Since both graphs show a roughly symmetrical distribution with no outliers, a t-test appears to be appropriate.

6.15 a. The data from the treatment group appears to be a random sample from a normal distribution but the data from the control group is definitely not normally distributed. The variance from the control group is 13.4 times larger than the variance from the treatment group. The Wilcoxon rank sum test is probably the most appropriate test statistic since it is more robust to deviations from stated conditions than is the separate variance t test.

b. The p-value from the separate variance t test is 0.053 and the p-value from the Wilcoxon test is 0.0438. With $\alpha = 0.05$, the conclusions from the two tests differ with the t test failing to find that the mean daily mileage for the treatment group significantly smaller than the mean for the control group. The Wilcoxon did find a significant reduction in the mean of the treatment group.

c. The t test is probably not appropriate since the control group appears to have a nonnormal distribution.

d. The choice of test statistic would definitely have an effect on the final conclusions of the study. When the test statistics yield contradictory results, and the required conditions of the test statistics are not met, it is best to error on the side of the most conservative conclusion relative to which of the two Types of errors has the greatest consequence. Hence, reject H_o only if one of the tests is highly significant relative to specified α.

6.17 a. To conduct the study using independent samples, the 30 participants should be very similar relative to age, body fat percentage, diet, and general health prior to the beginning of the study. The 30 participants would then be randomly assigned to the two treatments.

b. The participants should be matched to the greatest extend possible based on age, body fat, diet, and general health before the treatment is applied. Once the 15 pairs are configured, the two treatments are randomly assigned within each pair of participants.

c. If there is a large difference in the participants with respect to age, body fat, diet and general health and if the pairing results in a strong positive correlation in the responses from paired participants, then the paired procedure would be more effective. If the participants are quite similar in the desired characteristics prior to the beginning of the study, then the independent samples procedure would yield a test statistic having twice as many df as the paired procedure and hence would be more powerful.

6.19   a. $H_o : \mu_d = 0$ versus $H_a : \mu_d \neq 0$.

$t = 4.95, df = 29, \Rightarrow$ p-value $= 2P(t \geq 4.91) < 0.001$

There is significant evidence of a difference in the mean final grades.

b. A 95% confidence interval estimate of the mean difference in mean final grades is (2.23, 5.37).

c. We would need to verify that the difference in the grades between the 30 twins are independent. The normal probability plot would indicate that the differences are a random sample from a normal distribution. Thus, the conditions for using a paired t-test appear to be valid.

d. Yes. The purpose of pairing is to reduce the subject to subject variability and there appears to be considerable differences in the students in the study. Also, a scatterplot of the data yields a strong positive correlation between the scores for the twins. Scatterplot is given here:

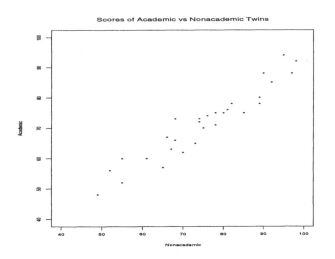

6.21 H_o : The distribution of differences (Benzedrine minus placebo) is symmetric about 0 versus

H_a : The differences tend to be larger than 0

With $n = 14, \alpha = 0.05, T = T_-$, reject H_o if $T_- \leq 25$.

From the data we obtain $T_- = 16 < 25$, thus reject H_o and conclude that the distribution of heart rates for dogs receiving Benzedrine is shifted to the right of the dogs receiving the placebo.

6.23 Let d=After-Before

    a. $H_o : \mu_d \leq 0$ versus $H_a : \mu_d > 0$;

       $t = \frac{1.208-0}{1.077/\sqrt{12}} = 3.89$, with $df = 11 \Rightarrow .001 < $ p-value $< 0.005 \Rightarrow$

       Reject H_o and conclude that the exposure has increased mean lung capacity.

    b. 95% C.I. on $\mu_{After} - \mu_{Before}$: (0.52, 1.89)

    c. If this was a well controlled experiment with all factors except ozone exposure controlled, the experimenter would be justified in making the claim concerning the population of rats from which the rats in the study were randomly selected. However, no statement can be made about the effects of ozone on humans.

6.25   a. $H_o : \mu_{Within} = \mu_{Out}$ versus $H_a : \mu_{Within} \neq \mu_{Out}$;

       The separate variance t-test: $t' = \frac{3092-2450}{\sqrt{\frac{(1191)^2}{14} + \frac{(2229)^2}{12}}} = 0.89 \Rightarrow$

       with $df \approx 16$, p-value $\approx 0.384 \Rightarrow$

       Fail to reject H_o and conclude the data does not provide sufficient evidence that there is a difference in average population abundance.

    b. A 95% C.I. on $\mu_{Within} - \mu_{Out}$ is (-879, 2164)

    c. The two samples are independently selected random samples from two normally distributed populations.

    d. Box plots are given here:

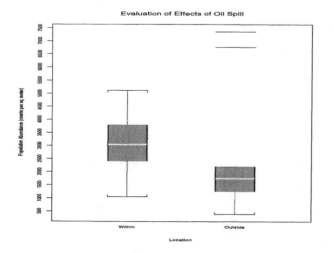

The Within data set appears to be normally distributed but the Outside data may not be normally distributed since there were two outliers. The sample variance of the Outside is 3.5 times larger than the Within sample variance.

6.27   a. Box plot is given here:

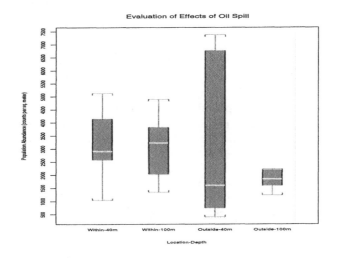

b. 40M: $H_o : \mu_{Within} = \mu_{Out}$ versus $H_a : \mu_{Within} \neq \mu_{Out}$;

The separate variance t-test: $t' = \dfrac{3183-3080}{\sqrt{\frac{(1289)^2}{7} + \frac{(3135)^2}{6}}} = 0.08 \Rightarrow$

with $df \approx 6$, p-value $\approx 0.94 \Rightarrow$

fail to reject H_o and conclude the data does not provide sufficient evidence that there is a difference in average population abundance at 40 m.

100M: $H_o : \mu_{Within} = \mu_{Out}$ versus $H_a : \mu_{Within} \neq \mu_{Out}$;

The separate variance t-test: $t' = \dfrac{3002-1820}{\sqrt{\frac{(1181)^2}{7} + \frac{(385)^2}{6}}} = 2.50 \Rightarrow$

with $df \approx 7$, p-value $\approx 0.041 \Rightarrow$

reject H_o and conclude the data provides sufficient evidence that there is a difference in average population abundance at 100 m.

c. The conclusions are different at the two depths. The mean population abundances are fairly consistent at all but the 100 m depth outside the oil trajectory where the abundance is considerably smaller. However, the median at both within depths are nearly the same but at a higher level than the medians at two outside depths.

6.29   a. $H_o : \mu_{High} = \mu_{Con}$ versus $H_a : \mu_{High} \neq \mu_{Con}$;

Separate variance t-test: $t' = 4.12$ with $df \approx 34$, p-value $= 0.0002. \Rightarrow$

Reject H_o and conclude there is significant evidence of a difference in the mean drop in blood pressure between the high-dose and control groups.

b. 95% C.I. on $\mu_{High} - \mu_{Con}$: $(19.5, 57.6)$, i.e., the high dose groups mean drop in blood pressure was, with 95% confidence, 19.5 to 57.6 points greater than the mean drop observed in the control group.

c. Provided the researcher independently selected the two random samples of participants, the conditions for using a separate variance t-test were satisfied since the plots do not detect a departure from a normally distribution but the sample variances are somewhat different(1.9 to 1 ratio).

31

6.31 a. $H_o : \mu_{Low} = \mu_{High}$ versus $H_a : \mu_{Low} \neq \mu_{High}$;

Separate variance t-test: $t = -5.73$ with $df \approx 29$, p-value < 0.0005. $\Rightarrow$

Reject H_o and conclude there is significant evidence of a difference in the mean drop in blood pressure between the high-dose and low-dose groups.

b. 95% C.I. on $\mu_{Low} - \mu_{High}$: (-87.6, -41.5), i.e., the low-dose group mean drop in blood pressure was, with 95% confidence, 41.5 to 87.6 points lower than the mean drop observed in the high-dose group.

c. Provided the researcher independently selected the two random samples of participants, the conditions for using a pooled t-test were satisfied since the plots do not detect a departure from a normally distribution and the sample variances are somewhat different(3.2 to 1 ratio).

6.33 a. $H_o : \mu_D \leq \mu_{RN}$ versus $H_a : \mu_D > \mu_{RN}$;

Separate variance t-test: $t = 9.04$ with $df \approx 76$, p-value < 0.0001 $\Rightarrow$

Reject H_o and conclude there is significant evidence that the mean score of the Degreed nurses is higher than the mean scores of the RN nurses.

b. p-value < 0.0001

c. 95% C.I. on $\mu_D - \mu_{RN}$: (35.3, 55.2) points

d. Since 40 is contained in the 95% C.I., there is a possibility that the difference in mean scores is less than 40. Hence, the differences observed may not be meaningful.

6.35 The required conditions are that the two samples are independently selected from populations having normal distributions with equal variances. The box plots do not reveal any indication that the population distributions were not normal. The sample variances have a ratio of 1.4 to 1.0, thus there is very little indication that the population variances were unequal.

6.37 $H_o : \mu_{SUV} \leq \mu_{Mid}$ versus $H_a : \mu_{SUV} > \mu_{Mid}$

a. The box plots do not look particularly like samples from normal distributions, both indicate that the distributions are skewed to the right. The ratio of sample variances is 2.5 to 1, which is within acceptable limits for using the pooled t-test (provided both distributions are normally distributed). The lack of normality with such small sample sizes would invalidate the use of the t-test.

b. With t-test, we obtain p-value $= 0.11$. This would indicate that there is not significant evidence that mean damage is greater for the SUV's.

With the Wilcoxon rank sum test, we obtain p-value $= 0.0203$. This would indicate that there is significant evidence that mean damage is greater for the SUV's.

c. The Wilcoxon rank sum test is more appropriate because the populations appear to be nonnormally distributed. This would make the results of the t-test questionable, especially with the sample sizes being quite small.

d. The Wilcoxon and t-test tests yield contradictory results. The Wilcoxon rank sum test is insensitive to extreme values whereas the t-test can be invalidated when the distributions are highly skewed.

6.39 The concept of statistical significance involves inherent uncertainty since it is based on a probability. Because evidence in legal cases is admissible only if it can be asserted with near certainty, any statistical evidence will only be useful if its associated probability approaches 1. Common practice uses 95% significance as meaning near certainty and the legal profession accepts this figure as a scientific consensus.

6.41   a.   1. The alternative hypothesis is that the mean lead level is greater in the Exposed children than in the Control children. Therefore, the null hypothesis is that the mean lead level in the Exposed children is less than or equal to the mean lead level in the Control children. In symbols, we have

$$H_o : \mu_D \leq 0 \text{ versus } H_a : \mu_D > 0,$$

where $D =$ Exposed - Control. The p-value for the 1-sided test is given by p-value $= Pr[t_{32} \geq 5.78] = .00000103$. Thus, there is significant evidence that the mean difference in the lead levels between children in the Exposed and Control groups is greater than 0. The p-value for the 1-sided test is 1/2 of the p-value for the 2-sided test.

2. If the lead-using factory had implemented a safety program to warn its employees about the possible sourses of lead outside the factory, then the employees could educate their children how to avoid lead exposure and hence have lower levels than children of non-employees.

b. A Wilcoxon signed rank test and the sign test yield a p-value less than .000 and hence agree with the t-test.

```
Wilcoxon Signed Rank Test: Exposed-Control

Test of median = 0.000000 versus median  >  0.000000

                    N for   Wilcoxon            Estimated
              N     Test    Statistic      P      Median
Exposed      33      33       561.0     0.000      30.50
Control      33                                    16.00


Sign Test for Median: Exposed-Control

Sign test of median = 0.00000 versus  >  0.00000

              N  Below  Equal  Above      P     Median
Exposed      33    0      0      33    0.0000    34.00
Control      33                                  16.00
```

Chapter 7: Inferences about Population Variances

7.1 a. 0.01

    b. 0.90

    c. $1 - 0.99 = 0.01$

    d. $1 - 0.01 - 0.01 = 0.98$

7.3 a. Let y be the quantity in a randomly selected jar:

    Proportion $= P(y < 32) = P(z < \frac{32-32.3}{.15}) = 0.0228 \Rightarrow 2.28\%$

    b. The plot indicates that the distribution is approximately normal because the data values are reasonably close to the straight-line.

    c. 95% C.I. on σ : $\left(\sqrt{\frac{(50-1)(.135)^2}{70.22}}, \sqrt{\frac{(50-1)(.135)^2}{31.55}} \right) \Rightarrow (0.113, 0.168)$

    d. $H_o : \sigma \leq 0.15$ versus $H_a : \sigma > 0.15$

    Reject H_o if $\frac{(n-1)(s)^2}{(.15)^2} \geq 66.34$

    $\frac{(50-1)(.135)^2}{(.15)^2} = 39.69 < 66.34 \Rightarrow$

    Fail to reject H_o and conclude the data does not support σ greater than 0.15.

    e. p-value $= P(\frac{(n-1)(s)^2}{(.15)^2} \geq 39.69)$

    Using the Chi-square tables with $df = 49, 0.10 <$ p-value < 0.90

    (Using a computer program, p-value $= 0.8262$).

7.5 a. The box plot is symmetric but there are four outliers. Since the sample size is 150, a few outliers would be expected. However, four out of 150 may indicate the population distribution may have heavier tails than a normal distribution. This may cause the values of s to be inflated.

    b. 99% C.I. on σ : $\left(\sqrt{\frac{(150-1)(9.537)^2}{197.21}}, \sqrt{\frac{(150-1)(9.537)^2}{108.29}} \right) \Rightarrow (8.290, 11.187)$

    c. $H_o : \sigma^2 \leq 90$ versus $H_a : \sigma^2 > 90$

    With $\alpha = 0.05$, reject H_o if $\frac{(n-1)(s)^2}{90} \geq 178.49$

    $\frac{(150-1)(9.537)^2}{90} = 150.58 < 178.49 \Rightarrow$

    Fail to reject H_o and conclude the data fails to support the statement that σ^2 is greater than 90.

7.7 a. The box plot is symmetric with a single outlier. Since the sample size is 81, a few outliers would be expected. Thus, the normality of the population distribution appears to be satisfied.

    b. $H_o : \sigma \geq 2$ versus $H_a : \sigma < 2$

    With $\alpha = 0.05$, reject H_o if $\frac{(n-1)(s)^2}{(2)^2} \leq 60.39$

$$\frac{(81-1)(1.771)^2}{(2)^2} = 62.73 > 60.39 \Rightarrow$$

Fail to reject H_o and conclude the data fails to support the contention that σ is less than 2. p-value = 0.0772

c. 95% C.I. on σ : $\left(\sqrt{\frac{(81-1)(1.771)^2}{106.63}}, \sqrt{\frac{(81-1)(1.771)^2}{57.15}}\right) \Rightarrow (1.534, 2.095)$

7.9 $H_o : \sigma_A^2 \le \sigma_B^2$ versus $H_a : \sigma_A^2 > \sigma_B^2$

With $\alpha = 0.05$, reject H_o if $\frac{s_A^2}{s_B^2} \ge 3.79$

$s_A^2/s_B^2 = 3.15 < 3.79 \Rightarrow$

Fail to reject H_o and conclude the data does not support σ_A^2 being greater than σ_B^2.

7.11 95% C.I. on $\sigma_{Comp.}$: $\left(\sqrt{\frac{(91-1)(53.77)^2}{118.14}}, \sqrt{\frac{(61-1)(53.77)^2}{65.65}}\right) \Rightarrow (46.93, 62.96)$

95% C.I. on $\sigma_{Conv.}$: $\left(\sqrt{\frac{(91-1)(36.94)^2}{118.14}}, \sqrt{\frac{(91-1)(36.94)^2}{65.65}}\right) \Rightarrow (32.24, 43.25)$

95% C.I. on $\mu_{Comp.}$: $484.45 \pm (1.987)(53.77)/\sqrt{91} \Rightarrow (473.60, 495.30)$

95% C.I. on $\mu_{Conv.}$: $487.38 \pm (1.987)(36.94)/\sqrt{91} \Rightarrow (479.69, 495.07)$

$H_o : \sigma_{Comp.}^2 = \sigma_{Conv.}^2$ versus $H_a : \sigma_{Comp.}^2 \ne \sigma_{Conv.}^2$.

With $\alpha = 0.05$, reject H_o if $\frac{s_{Comp.}^2}{s_{Conv.}^2} \le \frac{1}{1.52} = 0.660$ or $\frac{s_{Comp.}^2}{s_{Conv.}^2} \ge 1.52$

$s_{Old}^2/s_{New}^2 = 2.12 > 1.52 \Rightarrow$

Reject H_o and conclude there is significant evidence that $\sigma_{Comp.}^2$ and $\sigma_{Conv.}^2$ are different.

$H_o : \mu_{Comp.} = \mu_{Conv.}$ versus $H_a : \mu_{Comp.} \ne \mu_{Conv.}$.

$t = \frac{484.45 - 487.38}{\sqrt{\frac{(53.77)^2}{91} + \frac{(36.94)^2}{91}}} = -0.06 \Rightarrow$ p-value $= 2P(t \ge |-0.06|) = 0.95$

Fail to reject H_o and conclude there is not significant evidence that the mean SAT math exam scores are different.

The two methods yield similar mean scores but the computer testing method has a higher degree of variability than the conventional method.

7.13 The data is summarized in the following table:

Method	n	Mean	95% C.I. on μ	St.Dev.	95% C.I. on σ
L	10	5.90	(2.34, 9.46)	4.9766	(3.42, 9.09)
L/R	7	7.29	(2.31, 12.26)	5.3763	(3.46, 11.84)
L/C	9	16.00	(9.57, 22.43)	8.3666	(5.65, 16.03)
C	9	17.67	(5.45, 29.88)	15.8902	(10.73, 30.44)

From the box plots, it appears that the L/R distribution is right skewed but the other 3 distributions appear to be random samples from normal distributions.

Test $H_o : \sigma_L = \sigma_{L/R} = \sigma_{L/C} = \sigma_C$ versus $H_a : \sigma'$s are different

Reject H_o at level $\alpha = 0.05$ if $L \geq F_{.05,3,31} = 2.91$

From the data, $L = 2.345 < 2.91 \Rightarrow$

Fail to reject H_o and conclude there is not significant evidence of difference in variability of the increase in test scores.

Based on the C.I.'s for the $\mu's$, we can conclude that there is very little difference in the average change in test scores for the four methods of instruction. However, lecture only method yielded somewhat smaller mean change in test score than the computer instruction only procedure. These confidence intervals have an overall level of confidence of $(.95)^4 = 0.81$ since the data from the four procedures are independent. Thus, our conclusion would have a relatively large chance of committing a Type I error in attempting to determine if any pair of instructional methods have different means. An improved procedure for comparing the four instructional methods will be covered in Chapter 8. This procedure would determine that there is a significant difference in the instructional means (p-value = 0.032).

7.15   a. 25x90% = 22.5 and 25x110% = 27.5 implies the limits are 22.5 to 27.5

   b. The box plot and normal probability plot are given here:

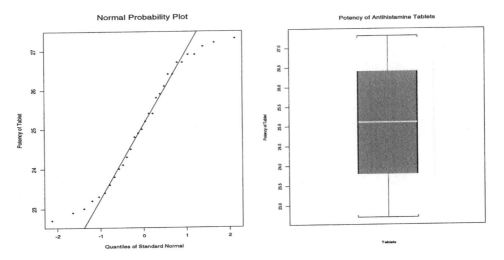

The box plot indicates a symmetric distribution with no outliers. The normal probability plot shows the data values reasonably close to a straight line, although there is some deviation at both ends which indicates that the data may be a random sample from a distribution which has shorter tails than a normally distributed population.

   c. Range = 27.5-22.5 = 5 $\Rightarrow \hat{\sigma} = 5/4 = 1.25$

   $H_o : \sigma = 1.25$ versus $H_a : \sigma \neq 1.25$

   With $\alpha = 0.05$, reject H_o if $\frac{(n-1)s^2}{(1.25)^2} \leq 16.05$ or $\frac{(n-1)s^2}{(1.25)^2} \geq 45.72$

   $\frac{(30-1)(1.4691)^2}{(1.25)^2} = 40.06 \Rightarrow 16.06 < 40.06 < 45.72$

   Fail to reject H_o and conclude there is insufficient evidence that the product standard deviation is greater than 1.25. Thus, it appears that the potencies are within the required bounds.

7.17 Since the box plots indicate that data from both portfolios has a normal distribution. Also, the C.I. on the ratio of the variances contained 1 which indicates equal variances. Thus, a pooled variance t-test will be used as the test statistic.

$H_o : \mu_1 = \mu_2$ versus $H_a : \mu_1 \neq \mu_2$

$t = \frac{131.60 - 147.20}{4.92 \sqrt{\frac{1}{10} + \frac{1}{10}}} = -7.09$ with df=18 $\Rightarrow$ p-value < 0.0005

Reject H_o and conclude that the data strongly supports a difference in the mean returns of the two portfolios.

7.19 We would now run a 1-tail test:

$H_o : \mu_A \geq \mu_B$ versus $H_a : \mu_A < \mu_B$

$t = \frac{27.62 - 34.69}{\sqrt{\frac{(9.83)^2}{13} + \frac{(4.03)^2}{13}}} = -2.40$ with df=15 $\Rightarrow$ p-value $= 0.0149$

Reject H_o and conclude that the data indicates the mean length of time people remain on therapy B is longer than the mean for therapy A.

Chapter 8: The Completely Randomized Design

8.1 a. Yes, the mean for Device A is considerably (relative to the standard deviations) smaller than the mean for Device D.

b. $H_o : \mu_A = \mu_B = \mu_C = \mu_D$ versus H_a : Difference in $\mu's$

Reject H_o if $F \geq F_{.05,3,20} = 3.10$

$SSW = 5[(.1767)^2 + (.2091)^2 + (.1532)^2 + (.2492)^2] = 0.8026$

$\bar{y}_{..} = 0.0826 \Rightarrow$

$SSB = 6[(-0.1605 - .0826)^2 + (0.0947 - .0826)^2 + (0.1227 - .0826)^2 + (0.2735 - .0826)^2]$

$\qquad = 0.5838 \Rightarrow$

$F = \frac{.5838/3}{.8026/20} = 4.85 > 3.10 \Rightarrow$

Reject H_o and conclude there is significant difference among the mean difference in pH readings for the four devices.

c. p-value $= P(F_{3,20} \geq 4.85) \Rightarrow$ p-value $= 0.0107$

d. The data must be independently selected random samples from normal populations having the same value for σ.

e. Suppose the devices are more accurate at higher levels of pH in the soil, and if by chance all soil samples with high levels of pH are assigned to a particular device, then that device may be evaluated as more accurate based just on the chance selection of soil samples and not on a true comparison with the other devices.

8.3 a. The box plot and normal probability plot are given here:

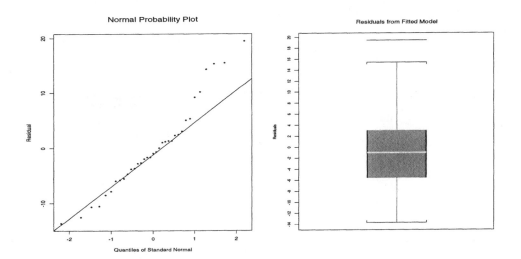

Based on the box plots and normal probability plots it appears that the distributions of several of the varities have a nonnormal distribution. The Levine's test yields L = 2.381 with p-value = 0.74. Thus, there is not a significant difference in the variety variances.

b. The AOV table is given here:

Source	df	SS	MS	F	p-value
Variety	4	1096.7	274.2	3.73	0.014
Error	30	2205.4	73.5		
Total	34	3302.1			

Reject H_o if $F \geq 2.69 \Rightarrow$

Since $F = 3.73 > 2.69$, reject H_o and conclude there is a significant difference in the mean yield of five varieties.

c. The Kruskal-Wallis yields $H' = 10.01$ with p-value = 0.040. Thus, reject H_o and conclude there is a significant difference in the distributions for the yields of the five varieties.

d. The only difference in the conclusions is that the p-value for the F-test is somewhat smaller than the p-value for the Kruskal-Wallis test.

8.5   a. The Kruskal-Wallis yields $H = 21.32 > 9.21$ with $df = 2 \Rightarrow$ p-value < 0.001. Thus, reject H_o and conclude there is a significant difference in the distributions of deviations for the three suppliers.

b. The box plots and normal probability plot are given here:

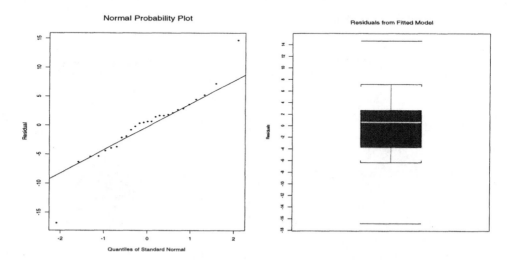

The box plots and normal probability plots of the residuals indicate that the normality condition may be violated. The Levine test yields $L = 3.89$ with $0.025 <$ p-value < 0.05 . Thus, there is significant evidence that the equal variance condition is also violated

c. The AOV table is given here:

Source	df	SS	MS	F	p-value
Supplier	2	10723.8	5361.9	161.09	0.000
Error	24	798.9	33.3		
Total	26	11522.7			

Reject H_o if $F \geq 3.40$

$F = 161.09 > 3.40$, reject H_o and conclude there is a significant difference in the mean deviations of the three suppliers.

d. 95% C.I. on μ_A: $189.23 \pm (2.064)(5.77)/\sqrt{9} \Rightarrow (185.26, 193.20)$

95% C.I. on μ_B: $156.28 \pm (2.064)(5.77)/\sqrt{9} \Rightarrow (152.31, 160.25)$

95% C.I. on μ_C: $203.94 \pm (2.064)(5.77)/\sqrt{9} \Rightarrow (199.97, 207.91)$

Since the upper bound on the mean for supplier B is more than 20 units less than the lower bound on the mean for suppliers A and B, there appears to be a practical difference in the three suppliers. However, because the normality and equal variance assumptions may not be valid, the C.I.'s may not be accurate.

8.7 The box plots indicate the distribution of the residuals is slightly right skewed. This is confirmed with an examination of the normal probability plot. The Hartley test yields $F_{max} = 2.35 < 7.11$ using an $\alpha = 0.05$ test. Thus, the conditions needed to run the ANOVA F-test appear to be satisfied. From the output, $F = 15.68$, with p-value $< 0.0001 < 0.05$. Thus, we reject H_o and conclude there is significant evidence of a difference in the average weight loss obtained using the five different agents.

8.9 a. Based on the box plots and the normal probability plot, the condition of normality of the population distributions appears to be satisfied. The Hartley test yields:

$F_{max} = \frac{(1.0670)^2}{(0.8452)^2} = 1.59 < 14.5$ (value from Table 12 with $\alpha = .01$) $\Rightarrow$ There is not significant evidence of a difference in the 4 population variances.

b. From the ANOVA table, we have p-value < 0.001. Thus, there is significant evidence that the mean ratings differ for the four groups.

c. 95% C.I. on μ_I : $8.3125 \pm 2.048 \frac{\sqrt{0.9763}}{\sqrt{8}} = (7.6, 9.0)$

95% C.I. on μ_{II} : $6.4375 \pm 2.048 \frac{\sqrt{0.9763}}{\sqrt{8}} = (5.7, 7.1)$

95% C.I. on μ_{III} : $4.0000 \pm 2.048 \frac{\sqrt{0.9763}}{\sqrt{8}} = (3.3, 4.7)$

95% C.I. on μ_{IV} : $2.5000 \pm 2.048 \frac{\sqrt{0.9763}}{\sqrt{8}} = (1.8, 3.2)$

d. The C.I.'s are the same as those given in the output.

8.11 a. The Kruskal-Wallis test yields: $H' = 26.62$ with df $= 3 \Rightarrow$ p-value $< 0.0001 \Rightarrow$ Thus, there is significant evidence that the distribution of ratings differ for the four groups.

b. The two procedures yield equivalent conclusions.

8.13   a. $F = \frac{4020.0/3}{881.9/36} = 54.70$ with df $= 3,36 \Rightarrow$ p-value $< 0.0001 < 0.05 \Rightarrow$

There is significant evidence of a difference in the average leaf size under the four growing conditions.

b. 95% C.I. on $\mu_A : 23.37 \pm 2.028\frac{\sqrt{881.9/36}}{\sqrt{10}} = (20.20, 26.54)$

95% C.I. on $\mu_B : 8.58 \pm 2.028\frac{\sqrt{881.9/36}}{\sqrt{10}} = (5.41, 11.75)$

95% C.I. on $\mu_C : 14.93 \pm 2.028\frac{\sqrt{881.9/36}}{\sqrt{10}} = (11.76, 18.10)$

95% C.I. on $\mu_D : 35.35 \pm 2.028\frac{\sqrt{881.9/36}}{\sqrt{10}} = (32.18, 38.52)$

The C.I. for the mean leaf size for Condition D implies that the mean is much larger for Condition D than for the other three conditions.

c. $F = \frac{18.08/3}{103.17/36} = 2.10$ with df $= 3,36 \Rightarrow$ p-value $= 0.1174 \Rightarrow$

There is not significant evidence of a difference in the average nicotine content under the four growing conditions.

d. From the given data, it is not possible to conclude that the four growing conditions produce different average nicotine content.

e. No. If the testimony was supported by this experiment, then the test conducted in part (c) would have had the opposite conclusion.

8.15   a. Because the data appears to be nonnormal in distribution, the Levene test will be used.

L $= 0.84$ with $df = 2, 17 \Rightarrow$ p-value $= 0.4489 \Rightarrow$

There is not significant evidence of a difference in population variances.

b. No, because the distributions appear to be not normally distributed.

c. The data was transformed using the logarithmic transformation.

The data is still somewhat skewed right for Machine C but the ANOVA will be conducted anyways.

$F = \frac{9.848/2}{11.964/17} = 7.00$ with df $= 2,17 \Rightarrow$ p-value $= 0.0061 \Rightarrow$

There is significant evidence of a difference in the mean diameters of the three machines.

d. The analysis using the original data yield p-value $= 0.094$ which would not support a difference in the mean diameters.

e. She could have selected an equal number of observations from each machine. Equal sample sizes has a moderating influence on the unequal variances effects on the F-test.

8.17   a. The summary statistics for the four groups are given here.

Descriptive Statistics: ActiveEx, PassivEx, TestOnly, Control

Variable	N	Mean	Median	TrMean	StDev	SE Mean
ActiveEx	6	10.125	9.625	10.125	1.447	0.591
PassivEx	6	11.375	10.750	11.375	1.896	0.774
TestOnly	6	11.708	11.750	11.708	1.520	0.621
Control	5	12.350	12.000	12.350	0.962	0.430

b. Box plots are given here:

Age (months) at Which Child First Walked

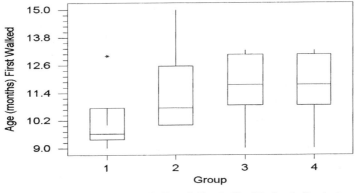

1=ActiveEx, 2=PassivEx, 3=TestOnly, 4=Control

c. The Control group has the largest mean age. The four groups have similar levels of varibility in ages with the Control groups standard deviation somewhat smaller than the other three groups. The box plots are not very informative because of the very small sample sizes in each of the four groups.

d. The F-test from the Minitab AOV has a p-value of 0.129 which indicates that there is not significant evidence in the data that the four groups have different mean ages.

e. The Minitab output for the LSD and Tukey procedure are given here.

```
Tukey's pairwise comparisons

    Family error rate = 0.0500
Individual error rate = 0.0111

Critical value = 3.98

Intervals for (column level mean) - (row level mean)

          ActiveEx    Control    PassivEx

Control     -4.809
             0.359

PassivEx    -3.714     -1.609
             1.214      3.559

TestOnly    -4.047     -1.942     -2.797
             0.881      3.226      2.131



Fisher's pairwise comparisons

Family error rate = 0.191
Individual error rate = 0.0500

Critical value = 2.093

Intervals for (column level mean) - (row level mean)

          ActiveEx    Control    PassivEx

Control     -4.147
            -0.303

PassivEx    -3.082     -0.947
             0.582      2.897

TestOnly    -3.416     -1.280     -2.166
             0.249      2.564      1.499
```

In Tukey's procedures, the 6 pairwise comparisons intervals all contain 0. Therefore, Tukey's procedure agrees with the conclusion from the F-test that there is not significant evidence of a difference in the four group means. Fisher's pairwise comparisons procedure has the interval for comparing the mean of the Control with the mean of ActiveEx group not containing 0 but with the other 5 pairwise comparisons intervals all containing 0. This would seem to contradict the F-test and Tukey conclusions. However, if we use the Protected Fisher LSD procedure, then no further comparisons would have been made once the conclusion of the F-test indicated no significant evidence of a difference. Thus, all three procedures, F-test, Tukey and Protected Fisher LSD, have reached the same conclusion: No significant evidence at the .05 level of a difference in the means of the four groups.

f. A plot of the average number of walking reflexes observed is given here for the three
groups.

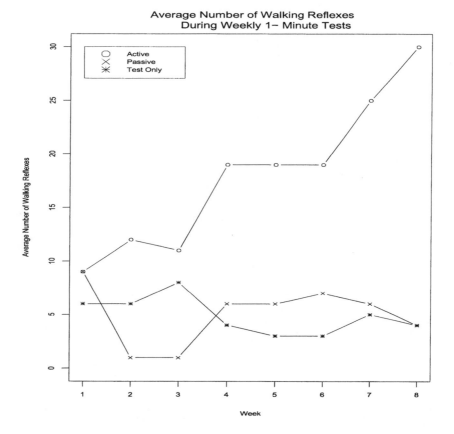

The average number of walking reflexes for the active exercise group increases dramatically over the 8 weeks whereas the values for the Passive and Test Only groups remains relatively constant.

Chapter 9: More Complicated Experimental Designs

9.1   a. The F-test from the ANOVA table tests 2-sided alternatives:

Test $H_o : \mu_{Attend} = \mu_{DidNot}$ vs $H_o : \mu_{Attend} \neq \mu_{DidNot}$

The ANOVA table is given here:

Source	DF	SS	MS	F	p-value
Pair	5	1319.42	263.88		
Treatment	1	420.08	420.08	71.40	0.0001
Error	5	29.42	5.88		
Total	11	1768.92			

Reject H_o and conclude there is significant evidence that the mean scores of students attending Head Start are significantly different from the mean scores of students who do not attend Head Start.

b. $RE(RCB,CR) = \frac{(b-1)MSB + b(t-1)MSE}{(bt-1)MSE} = \frac{(6-1)(263.88) + (6)(2-1)(5.88)}{((6)(2)-1)(5.88)} = 20.94 \Rightarrow$

It would take approximately 21 times as many observations (126) per treatment in a completely randomized design to achieve the same level of precision in estimating the treatment means as was accomplished in the randomized complete block design.

9.3   a. Blocks are Investigators and Treatments are Mixtures

b. Randomly assign the four Mixtures to the each of the Investigators

c. The randomized complete block design guarantees that each investigator measures each of the four mixtures, whereas in a completely randomized design, it is possible that some of the investigators may not measure some of the mixtures. This may cause a bias towards some of the mixtures if a particular investigator tends to always give high readings no matter which mixture is measured.

9.5   a. $y_{ij} = \mu + \alpha_i + \beta_j + \epsilon_{ij}; \quad i = 1, 2, 3, \quad j = 1, 2, 3, 4, 5, 6, 7$

y_{ij} is score on test of jth subject hearing the ith music type

α_i is the ith music type effect

β_j is the jth subject effect

$\hat{\mu} = 21.33, \quad \hat{\alpha}_1 = -0.47, \quad \hat{\alpha}_2 = -1.19, \quad \hat{\alpha}_3 = 1.67$

$\hat{\beta}_1 = 0, \quad \hat{\beta}_2 = -3, \quad \hat{\beta}_3 = 3.33, \quad \hat{\beta}_4 = -1.33, \quad \hat{\beta}_5 = 1, \quad \hat{\beta}_6 = 3.67, \quad \hat{\beta}_7 = -3.67$

b. $F = \frac{SS_{TRT}/df_{TRT}}{SS_{Error}/df_{Error}} = \frac{30.952/2}{28.38/12} = 6.54$ with df=2,12.

Therefore, p-value $= Pr(F_{2,12} \geq 6.54) = 0.0120 \Rightarrow$ Reject $H_o : \mu_1 = \mu_2 = \mu_3$.

We thus conclude that there is significant evidence of a difference in mean typing scores for the three types of music.

c. An interaction plot of the data is given here:

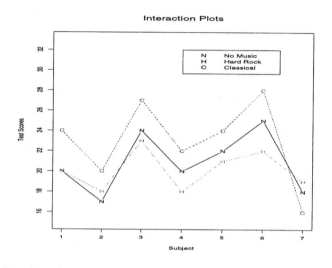

Interaction Plots

Based on the interaction plot, the additive model may be inappropriate because there is some crossing of the three lines. However, the plotted points are means of a single observation and hence may be quite variable in their estimation of the population means μ_{ij}. Thus, exact parallelism is not required in the data plots to ensure the validity of the additive model.

d. $t = 3, b = 7 \Rightarrow RE(RCB, CR) = \frac{(7-1)(24.889)+(7)(3-1)(2.365)}{((7)(3)-1)(2.365)} = 3.86 \Rightarrow$

It would take 3.86 times as many observations (approximately 27) per treatment in a completely randomized design to achieve the same level of precision in estimating the treatment means as was accomplished in the randomized complete block design. Since RE was much larger than 1, we would conclude that the blocking was effective.

9.7 a. $y_{ij} = \mu + \alpha_i + \beta_j + \epsilon_{ij}; \quad i = 1, 2, 3, 4 \quad j = 1, 2, 3, 4, 5$

y_{ij} is the increase in productivity of worker having jth level of attitude and attending workshop type ith

α_i is the ith workshop type effect

β_j is the jth attitude effect

$\hat{\mu} = 50.25, \quad \hat{\alpha}_1 = -7.45, \quad \hat{\alpha}_2 = -3.65, \quad \hat{\alpha}_3 = 0.35, \quad \alpha_4 = 10.75$

$\hat{\beta}_1 = -9.75, \quad \hat{\beta}_2 = -8.5, \quad \hat{\beta}_3 = -4.75, \quad \hat{\beta}_4 = 0.5, \quad \hat{\beta}_5 = 22.5$

b. $F = \frac{SS_{TRT}/df_{TRT}}{SS_{Error}/df_{Error}} = \frac{922.55/3}{55.7/12} = 114.12$ with df=3,12.

Therefore, p-value $= Pr(F_{3,12} \geq 114.12) < 0.0001 \Rightarrow$ Reject $H_o : \mu_1 = \mu_2 = \mu_3 = \mu_4$.

We thus conclude that there is significant evidence of a difference in the mean increase in productivity for the four types of workshops.

c. A profile plot of the data is given here:

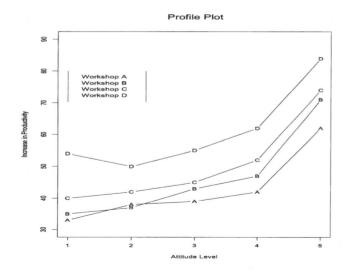

Profile Plot

Based on the profile plot, the additive model appears to be appropriate because the four lines are relatively parallel. Note further that the plotted points are means of a single observation and hence may be quite variable in their estimation of the population means μ_{ij}. Thus, exact parallelism is not required in the data plots to ensure the validity of the additive model.

d. $t = 4, b = 5 \Rightarrow RE(RCB, CR) = \frac{(5-1)(696.375)+(5)(4-1)(4.6417)}{((5)(4)-1)(4.6417)} = 32.37 \Rightarrow$

It would take 32.37 times as many observations (approximately 162) per treatment in a completely randomized design to achieve the same level of precision in estimating the treatment means as was accomplished in the randomized complete block design. Since RE was much larger than 1, we would conclude that the blocking was effective.

9.9   a. $y_{ijk} = \mu + \alpha_k + \beta_i + \gamma_j + \epsilon_{ijk}; \quad i, j, k = 1, 2, 3, 4;$

where y_{ijk} is the dry weight of a watermellon plant grown in Row i and Column j receiving Treatment k.

α_k is the effect of the kth Treatment on dry weight

β_i is the effect of the ith Row on dry weight

γ_j is the effect of the jth Column on dry weight

b. The Row, Column, and Treatment Means are given here:

Level	1	2	3	4
Row Mean $\bar{y}_{i..}$	1.53	1.5475	1.545	1.5475
Column Mean $\bar{y}_{.j.}$	1.5625	1.575	1.505	1.5275
Treatment Mean $\bar{y}_{..k}$	1.7375	1.685	1.4225	1.325

The overall mean is $\bar{y}_{...} = 1.5425$

The parameter estimates are given here:

$\hat{\mu} = 1.5425, \quad \hat{\beta}_1 = -.0125, \quad \hat{\beta}_2 = .005, \quad \hat{\beta}_3 = 0.0025, \quad \hat{\beta}_4 = .005$

49

$$F = \frac{SS_{Trt}/df_{Trt}}{SS_{Error}/df_{Error}} = \frac{.48015/3}{.00075/6} = 1280.4 \text{ with } df = 3,6 \Rightarrow \text{p-value} < 0.0001$$

Reject $H_o : \mu_1 = \mu_2 - \mu_3 = \mu_4$ and conclude there is significant evidence that the four treatments have different mean dry weights.

9.11   a. Similar results.

   b. The Boxplot and normal probability plot do not indicate a deviation from a normal distribution for the residuals.

   The plot of Residuals vs Pred do not indicate a deviation from the constant variance condition.

   Based on these plots, there does not indicate any deviations from the model conditions.

9.13   a. A profile plot of the data is given here:

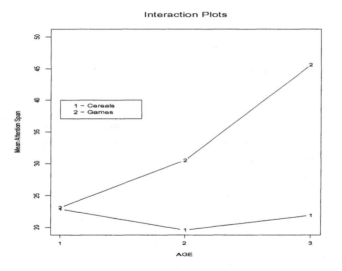

Interaction Plots

The profile plot indicates an increasing effect of Product Type as Age increases.

   b. The p-value for the interaction term is 0.013. There is significant evidence of an interaction between the factors Age and Product Type. Thus, the size of the difference between mean attention span of children viewing breakfast cereals and viewing video games would be different for the three age groups. From the profile plots, the estimated mean attention span for video games is larger than for breakfast cereals, with the size of the difference becoming larger as age increases.

9.15 There are 15 treatments consisting of the 3 levels of Factor A combined with the 5 levels of Factor B. These 15 treatments will be randomly assigned to 15 experimental units in each of the three blocks as seen in the following diagram:

	Block 1			Block 2			Block 3		
	Factor B			Factor B			Factor B		
Factor A	B1	B2	B3	B1	B2	B3	B1	B2	B3
A1	x	x	x	x	x	x	x	x	x
A2	x	x	x	x	x	x	x	x	x
A3	x	x	x	x	x	x	x	x	x
A4	x	x	x	x	x	x	x	x	x
A5	x	x	x	x	x	x	x	x	x

Source	DF	SS	MS	F	p-value
Treatment	14	SST	MST		
Factor A	2	SSA	MSA	MSA/MSE	
Factor B	4	SSB	MSB	MSB/MSE	
Interaction	8	SSAB	MSAB	MSAB/MSE	
Blocks	2	SSBL	MSBL		
Error	28	SSE	MSE		
Total	44	SSTOT			

9.17   a. This is a randomized complete block design with the eight regions serving as the blocking variable and type of job serving as the treatment.

$y_{ij} = \mu + \alpha_i + \beta_j + \epsilon_{ij}; \quad i = 1,2,3, \quad j = 1,2,3,4,5,6,7,8$

y_{ij} is starting salary in region j of job type i

α_i is the ith job type effect on starting salary

β_j is the jth region effect on starting salary

b. The Minitab output is given here:

```
General Linear Model: y versus Region, Group

Analysis of Variance for y, using Adjusted SS for Tests

Source    DF    Adj SS    Adj MS      F      P
Region     7    42.620     6.089  14.42  0.000
Group      2    79.491    39.745  94.16  0.000
Error     14     5.909     0.422
Total     23   128.020
```

Because the p-value $< 0.0001 < 0.05 = \alpha$, we can conclude there is significant evidence that the mean starting salary for the three groups of employees is different.

c. p-value < 0.0001

d. Using Tukey's W-procedure with $\alpha = 0.05$, $s_\epsilon^2 = MSE = .422$, $q_\alpha(t, df_{error}) = q_{.05}(3, 14) = 3.70 \Rightarrow$

$W = (3.70)\sqrt{\frac{.422}{8}} = .85 \Rightarrow$

Employee Group	Inspectors	Policemen	Firemen
Sample Mean	26.68	30.42	30.64
Tukey Grouping	a	b	b

Thus, Inspectors have significantly lower starting salaries than both Policemen and Firemen.

9.19 a. The profile plot is given here:

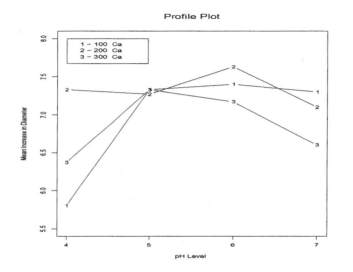

There appears to be an interaction between Ca Rate and pH with respect to the increase in trunk diameters. At low pH value, a 300 level of Ca yields largest increase; whereas, at high pH value, a 100 level of Ca yields largest increase in trunk diameter.

b. A model for this experiment is given here:

$y_{ijk} = \mu + \alpha_i + \beta_j + \alpha\beta_{ij} + \epsilon_{ijk}$; $i = 1, 2, 3, 4$; $j = 1, 2, 3$; $k = 1, 2, 3$;

where y_{ijk} is the increase in trunk diameter of the kth tree in soil having the ith pH level using the jth Ca Rate:

α_i is the effect of the ith pH level on diameter increase

β_j is the effect of the jth Ca Rate on diameter increase

$\alpha\beta_{ij}$ is the interaction effect of the ith pH level and jth Ca Rate on diameter increase.

c. This is a completely randomized 4x3 factorial experiment with Factor A: pH level, Factor B: Ca rate. There are 3 complete replications of the experiment. The AOV table is given here:

Source	DF	SS	MS	F	p-value
pH	3	4.461	1.487	21.94	0.0001
Ca	2	1.467	0.734	10.82	0.0004
Interaction	6	3.255	0.543	8.00	0.0001
Error	24	1.627	0.0678		
Total	35	10.810			

9.21 a. Using Tukey's W-procedure with $\alpha = 0.05$, $s_\epsilon^2 = MSE = .0678$, $q_\alpha(t, df_{error}) = q_{.05}(3, 24) = 3.53 \Rightarrow$

$$W = (3.53)\sqrt{\frac{.0678}{3}} = 0.53 \Rightarrow$$

pH		Ca Rate		
		100	200	300
4	Mean	5.80	7.33	6.37
	Grouping	a	c	b
5	Mean	7.33	7.27	7.33
	Grouping	a	a	a
6	Mean	7.40	7.63	7.17
	Grouping	a	a	a
7	Mean	7.30	7.10	6.60
	Grouping	b	ab	a

b. From the above table we observe that at pH=5, 6 there is not significant evidence of a difference in mean increase in diameter between the three levels of Ca. However, at pH=4, 7 there is significant evidence of a difference with Ca=300 yielding the largest increase at pH=4 and Ca=100 or 200 yielding the largest increase at pH=7. This illustrates the interaction between Ca and pH, i.e., the size of differences in the means across the levels of Ca depend on the level of pH.

9.23 a. The design is a completely randomized 5x5 factorial experiment with 2 replications; Factor A is Exterior Temperature and Factor B is Pane Design.

A model for this experiment is given here:

$y_{ijk} = \mu + \alpha_i + \beta_j + \alpha\beta_{ij} + \epsilon_{ijk}$; $i = 1, 2, 3, 4, 5$; $j = 1, 2, 3, 4, 5$; $k = 1, 2$;
where y_{ijk} is the heat loss of the kth pane having the ith temperature level and the jth pane design:
α_i is the effect of the ith temperature on heat loss
β_j is the effect of the jth pane desing on heat loss
$\alpha\beta_{ij}$ is the interaction effect of the ith temperature and jth pane design on heat loss

53

b. The test for an interaction between exterior temperature and pane design yields p-value $= 0.0073$ which would indicate significant evidence that an interaction exists. Therefore, the differences in mean heat loss between the five pane designs varies depending on the exterior temperature. The test of the main effect of pane design is not informative due to the significant interaction.

c. No, because of the significant interaction between pane design and exterior temperature. A profile plot is given here along with a table of the treatment means:

Temperature	Pane Design				
	A	B	C	D	E
0	10.50	10.50	11.45	11.60	9.60
20	9.50	9.50	10.45	10.60	9.55
40	9.45	9.50	10.40	10.45	9.45
60	8.10	8.45	8.55	8.45	9.55
80	7.50	7.50	8.45	8.60	9.55

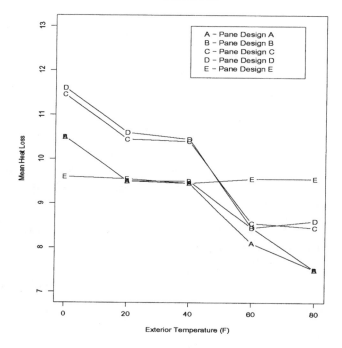

Profile Plot

d. Using Tukey's W-procedure with $\alpha = 0.05, s_\epsilon^2 = MSE = .2312, q_\alpha(t, df_{error}) = q_{.05}(5, 25) = 4.16 \Rightarrow$

$$W = (4.16)\sqrt{\frac{.2312}{2}} = 1.41 \Rightarrow$$

54

		Pane Design				
		A	B	C	D	E
Temp=0	Mean	10.50	10.50	11.45	11.60	9.60
	Grouping	ab	ab	b	b	a
Temp=20	Mean	9.50	9.50	10.45	10.60	9.55
	Grouping	a	a	a	a	a
Temp=40	Mean	9.45	9.50	10.40	10.45	9.45
	Grouping	a	a	a	a	a
Temp=60	Mean	8.10	8.45	8.55	8.45	9.55
	Grouping	a	a	a	a	b
Temp=80	Mean	7.50	7.50	8.45	8.60	9.55
	Grouping	a	a	a	a	b

From the above table we observe that at the exterior temperatures of $20°F$ and $40°F$ there is not significant evidence of a difference in mean heat loss between the five pane designs. However, at the exterior temperatures of $60°F$ and $80°F$ pane design E has a significantly higher mean heat loss than the other four designs. At exterior temperature of $0°F$, there are two groups of pane designs relative to their mean heat loss. This illustrates the interaction between exterior temperature and pane design, i.e., the size of differences in mean heat loss between the five pane designs depends on the exterior temperature.

9.25　a. The experiment is run as three reps of a completely randomized design with a 2x4 factorial treatment structure. A model for the experiment is given here:

$y_{ijk} = \mu + \alpha_i + \beta_j + \alpha\beta_{ij} + \epsilon_{ijk}; \quad i = 1, 2, 3, 4; \quad j = 1, 2; \quad k = 1, 2, 3;$

where y_{ijk} is the amount of active ingredient (or pH) of the kth vial having the ith storage time in laboratory j:

α_i is the effect of the ith storage time on amount of active ingredient (or pH)

β_j is the effect of the jth laboratory on amount of active ingredient (or pH)

$\alpha\beta_{ij}$ is the interaction effect of the ith storage time and jth laboratory on amount of active ingredient (or pH)

b. The complete AOV table is given here:

Source	DF	SS	MS	F	p-value
Storage Time	3	SSA	SSA/3	MSA/MSE	
Laboratory	1	SSB	SSB/1	MSB/MSE	
Interaction	3	SSAB	SSAB/3	MSAB/MSE	
Error	16	SSE	SSE/16		
Total	23	SST			

9.27   a. With respect to Amount of Active Ingredient:

The profile plots for the two-way interactions are given here:

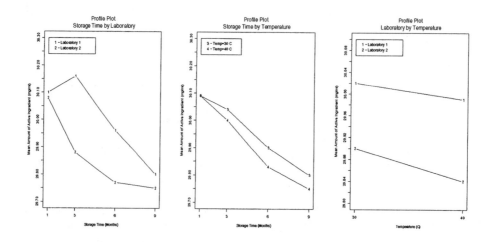

The Time*Lab*Temp interaction (p-value=.6914), Lab*Temp interaction (p-value=.3519), and Time*Temp interaction (p-value=.2817) were not significant. However, there was significant evidence (p-value=.0192) of a difference in the means for Amount of Active Ingredient at the two temperatures. There is a strong interaction between Time and Lab (p-value<.0001). Thus, comparisons of the means at the four storage times should be done separately for each Lab. The main effects of Lab and Time are not informative since these two factors have a significant interaction.

b. With respect to pH:

The profile plots for the two-way interactions are given here:

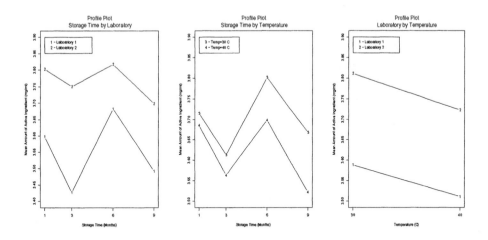

The Time*Lab*Temp interaction (p-value=.0621), Lab*Temp interaction (p-value = .7617), and Time*Temp interaction (p-value=.0686) were not significant. However, there was significant evidence (p-value < .0001) of a difference in the means for pH at the two temperatures. There is a strong interaction between Time and Lab (p-value=.0028). Thus, comparisons of the means at the four storage times should be done separately for each Lab. The main effects of Lab and Time are not informative since these two factors have a significant interaction.

c. Because the interaction between storage time and laboratory is significant, the effects of storage time differ between the two labs. Lab 1 has highest mean pH after 6 months, followed by months 1,9 and 3 months, respectively. In contrast, Lab 2 shows the highest mean pH for 6 months and 1 month, followed by 3 and 9 months, respectively.

Because the interaction between temperature and laboratory is not significant, the effects of temperature are the same for the two labs, with highest pH at 30°C.

d. Because the interaction between storage time and laboratory is significant, the effects of storage time differ between the two labs. Lab 1 has the highest mean active ingredient concentration after 3 months of storage time, followed by 1, 6, and 9 months, respectively. In contrast, Lab 2 shows its highest mean active ingredient concentration after 1 month of storage time, followed by 3, 6, and 9 months, respectively.

Because the interaction between temperature and laboratory is not significant, the effects of storage time are the same for the two labs, with highest pH at 30°C..

9.29 a. Latin Square Design with blocking variables Farm and Fertility. The treatment is the five types of fertilizers.

b. There is significant evidence (p-value < 0.0001) the mean yields are different for the five fertilizers.

Chapter 10: Categorical Data

10.1   a. Yes, because $n\hat{\pi} = 30 > 5$ and $n(1 - \hat{\pi}) = 120 > 5$. Samples with $n < 25$ would be suspect.

   b. $.2 \pm 1.645\sqrt{(.2)(.8)/150} \Rightarrow (0.15, 0.25)$ is a 90% C.I. for π.

10.3   a. $\hat{\pi} = 1202/1504 = 0.8 \Rightarrow$ 95% C.I. for $\pi : 0.8 \pm 1.96\sqrt{(.8)(.2)/1504} \Rightarrow (0.780, 0.820)$

   b. 90% C.I. for $\pi : 0.8 \pm 1.645\sqrt{(.8)(.2)/1500} \Rightarrow (0.783, 0.817)$

10.5 The 95% C.I.'s are summarized here: Note that $\hat{\pi}$ remains essentially unchanged from % Responding because n=1230 is very large.

Condition	95% C.I. on proportion having condition
Sore Throat	$0.30 \pm 1.96\sqrt{(.30)(.70)/1234} \Rightarrow (0.274, 0.326)$
Burns	$0.28 \pm 1.96\sqrt{(.28)(.72)/1234} \Rightarrow (0.255, 0.305)$
Alcohol	$0.25 \pm 1.96\sqrt{(.25)(.75)/1234} \Rightarrow (0.226, 0.274)$
Overweight	$0.22 \pm 1.96\sqrt{(.22)(.78)/1234} \Rightarrow (0.197, 0.243)$
Pain	$0.21 \pm 1.96\sqrt{(.21)(.79)/1234} \Rightarrow (0.187, 0.233)$

10.7 $\hat{\pi} = 88/254 = 0.346 \Rightarrow$ 90% C.I. for $\pi : 0.346 \pm 1.645\sqrt{(0.346)(0.654)/254} \Rightarrow (0.297, 0.395)$

10.9   a. A bar chart with the responses along the horizontal axis and the percentages along the vertical axis would allow comparison of the responses.

   b. Yes, since the C.I.'s would reflect the sampling errors of the point estimators and hence be more informative of the size of the true proportions.

   c. The report lists only a select few responses in the United States, ignoring the most and least popular ones as well as almost all of the foreign figures. For those percentages reported, it does not include the sample size and so the reader gets no idea of the accuracy of the reported sample proportions as estimates of the population proportions.

10.11   a. $n\pi_o = (800)(0.096) = 76.8 > 5$ and $n(1 - \pi_o) = (800)(1 - 0.096) = 723.2 > 5$ thus the normal approximation would be valid.

   b. $H_o : \pi \geq 0.096$ versus $H_a : \pi < 0.096$
   $\hat{\pi} = 35/800 = 0.04375, //z = \frac{0.04375 - 0.096}{\sqrt{(0.096)(0.904)/800}} = -5.02 \Rightarrow$
   p-value $= P(z < -5.02) < 0.0001 \Rightarrow$ Reject H_o and
   conclude there is significant evidence that $\pi < 0.096$.

10.13 $\hat{\pi} = 10/24 = 0.417 \Rightarrow$ 95% C.I. on $\pi : (0.220, 0.614)$

10.15 $\hat{\pi}_1 = 109/200 = 0.545$ (Republicans) and $\hat{\pi}_2 = 86/200 = 0.43$ (Democrats)

$z = \dfrac{0.545-0.43}{\sqrt{\frac{0.545(1-0.545)}{200} + \frac{0.43(1-0.43)}{200}}} = 2.32 \Rightarrow$ p-value $= 0.0102$

Reject H_o and conclude that a large proportion of Republicans are in favor of the incentives.

10.17 95% C.I. on $\pi_1 - \pi_2 : (0.013, 0.167)$

Because 0 is not contained within the C.I., H_o is rejected and hence the conclusion is the same as was in Exercise 10.16.

10.19 $H_o : \pi_1 = \pi_2$ versus $H_a : \pi_1 \neq \pi_2$

p-value $= 0.0018 < 0.05 \Rightarrow$ Reject H_o and conclude there is significant evidence that the population proportions are different.

10.21   a. $z = \dfrac{0.90-0.36}{\sqrt{\frac{0.9(1-0.9)}{100} + \frac{0.36(1-0.36)}{100}}} = 9.54 \Rightarrow$ p-value $= P(z > 9.54) < 0.0001$

    Reject H_o and conclude there is significant evidence that the death rate after 30 days is greater for Cocaine group than for the Heroin group.

  b. If the physical response to the two drugs is the same for humans, cocaine is a very dangerous drug, even more so than heroin.

10.23 $H_o : \pi_1 = \frac{1}{3}, \pi_2 = \frac{1}{3}, \pi_3 = \frac{1}{3}$

$H_a :$ at least on of the groups had probability of interning different from $\frac{1}{3}$

$E_i = n\pi_{io} = 63\pi_{io} \Rightarrow \quad E_1 = 21, \quad E_2 = 21 \quad E_3 = 21$

$\chi^2 = \sum_{i=1}^{3} \frac{(n_i - E_i)^2}{E_i} = 6.952$ with $df = 3 - 1 = 2 \Rightarrow$ p-value $= 0.0309 > 0.01 \Rightarrow$

Fail to reject H_o. The data does not appear to contradict the claim that students finishing an internship are equally distributed from the three industries.

10.25 $H_o : \pi_1 = 0.50, \pi_2 = 0.40, \pi_3 = 0.10$

$H_a :$ at least on of the $\pi_i s$ differs from its hypothesized value

$E_i = n\pi_{io} \Rightarrow \quad E_1 = 200(.5) = 100, \quad E_2 = 200(.4) = 80, \quad E_3 = 200(.1) = 20$

$\chi^2 = \sum_{i=1}^{3} \frac{(n_i - E_i)^2}{E_i} = 6.0$ with $df = 3 - 1 = 2 \Rightarrow$ p-value $= 0.0498 \Rightarrow$

Reject H_o at the $\alpha = 0.05$ level. There is substantial evidence that the distribution of registered voters is different from previous elections.

10.27 $H_o : \pi_1 = 0.0625, \pi_2 = 0.25, \pi_3 = 0.375, \pi_4 = 0.25, \pi_5 = 0.0625$

$H_a :$ at least on of the $\pi_i s$ differs from its hypothesized value

$E_i = n\pi_{io} \Rightarrow \quad E_1 = 125(.0625) = 7.8125, \quad E_2 = 125(.25) = 31.25,$

$E_3 = 125(.375) = 46.875, \quad E_4 = 125(.25) = 31.25, \quad E_5 = 125(.0625) = 7.8125$

$\chi^2 = \sum_{i=1}^{5} \frac{(n_i - E_i)^2}{E_i} = 7.608$ with $df = 5 - 1 = 4 \Rightarrow$ p-value $= 0.1070 \Rightarrow$

Fail to reject H_o. The data appear to fit the hypothesized theory that the securities analysts perform no better than chance, however, we have no indication of the probability of a Type II error.

10.29 Yes, since the row percentages differ considerably for the four categories of schools.

10.31   p-value = 0.0046

10.33   a. The 25%, 40%, and 35% claims concerning opinions on union membership was made for industrial workers as a whole without regard to membership status. The relevant data are the column totals of those favoring, those indifferent, and those opposed industrial workers, i.e., 210, 240, and 150, respectively.

b. The following table summarizes the information needed for the goodness-of-fit test:

Preference	Theoretical Proportions π_i	Expected Frequencies $E_i = 600\pi_i$	Observed Frequencies n_i
Favor	0.25	150	210
Indifferent	0.40	240	240
Oppose	0.35	210	150

$H_o : \pi_1 = 0.25, \pi_2 = 0.40, \pi_3 = 0.35$ versus H_a : Specified proportions are not correct
$\chi^2 = \sum_{i=1}^{3} \frac{(n_i - E_i)^2}{E_i} = 41.143$ with $df = 3 - 1 = 2 \Rightarrow$ p-value $< 0.0001 \Rightarrow$
Reject H_o. There is significant evidence that the speaker's claim is not supported by the data.

10.35 H_o : Membership Status and Opinion are independent Versus H_a : Membership Status and Opinion are related

The expected values in each cell and the cell chi-square values are given in the following table with the expected given above the cell chi-square values:

Status	Favor	Indifferent	Opposed
Members	70	80	50
	70.00	18.05	20.48
Nonmembers	140	160	100
	35.00	9.03	10.24

$\chi^2 = \sum_{i,j} \frac{(n_{ij} - E_{ij})^2}{E_{ij}} = 162.80$ with $df = (2-1)(3-1) = 2 \Rightarrow$ p-value $< 0.0001 \Rightarrow$
Reject H_o. There is significant evidence that the Membership Status and Opinion are related.

10.37   a. Under the hypothesis of independence, the expected frequencies are given in the following table: $\hat{E}_{ij} = n_{i.}n_{.j}/900$

	Opinion				
Commercial	1	2	3	4	5
A	42	107	78	34	39
B	42	107	78	34	39
C	42	107	78	34	39

b. df = (3-1)(5-1) = 8

c. The cell chi-squares are given in the following table:

	Opinion				
Commercial	1	2	3	4	5
A	2.3810	3.7383	2.1667	4.2353	0.6410
B	2.8810	10.8037	0.0513	5.7647	21.5641
C	0.0238	1.8318	1.5513	0.1176	14.7692

$\chi^2 = \sum_{i,j} \frac{(n_{ij} - E_{ij})^2}{E_{ij}} = 72.521$ with $df = 8 \Rightarrow$ p-value $< 0.0001 \Rightarrow$
Reject H_o. There is significant evidence that the Commercial viewed and Opinion are related.

10.39   a. The null hypothesis of statistical independence is equivalent to stating that there is no relationship between the type of order form and whether or not an order is received.

b. The p-value = 0.00007 from the chi-square test of independence. Thus, we reject the null hypothesis of independence and conclude that there is an association between the type of order form and whether or not an order was received.

10.41   a. Control: 10%; Low Dose: 14%; High Dose: 19%

b. $H_o : \pi_1 = \pi_2 = \pi_3$ versus H_a : The proportions are not all equal, where π_j is probability of a rat in Group j having One or More Tumors.

$E_{ij} = 100 n._j / 300$ and $\chi^2 = \sum_{ij} \frac{(n_{ij} - E_{ij})^2}{E_{ij}} = 3.312$ with df = (2-1)(3-1) = 2 and p-value = 0.191.

Because the p-value is fairly large, we fail to reject H_o and conclude there is not significant evidence of a difference in the probability of having One or More Tumors for the three rat groups.

c. No, since we the chi-square test failed to reject H_o.

10.43   a. Expected cell counts are given here:

	Years of Education			
Years First Job	0-4.5	4.5-9	9-13.5	> 13.5
0-2.5	17.90	20.97	23.78	26.34
2.5-5	24.14	28.28	32.07	35.52
5-7.5	16.70	19.56	22.18	24.57
> 7.5	11.26	13.20	14.97	16.57

b. $\chi^2 = 57.830$, with df $= (4\text{-}1)(4\text{-}1) = 9$

c. p-value < 0.001

d. There is significant evidence that Years of Education are related to Years on First Job.

10.45 a. Building: $\chi^2 = 34.167$ with p-value < 0.0001. Thus, there is substantial evidence that the customers from the different Ownership Group categories have different distributions relative to the Building Ratings categories.

Service: $\chi^2 = 18.117$ with p-value $= 0.112$. Thus, there is not significant evidence that customers from the different Ownership Group categories have different different distributions relative to the Service Ratings categories.

Chapter 11: Linear Regression and Correlation

11.1 A scatterplot of the data is given here:

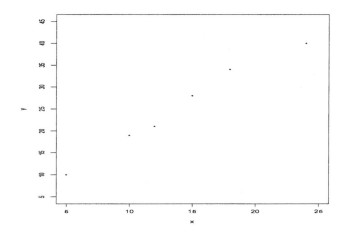

11.3 a. $\hat{y} = 4.698 + 1.97x$

b. yes, the plotted line is relatively close to all 10 data points

c. $\hat{y} = 4.698 + (1.97)(35) = 73.65$

11.5 $\hat{y} = 48.935 + 10.33(1.0) = 59.265$

11.7 a. For the transformed data, the plotted points appear to be reasonably linear.

b. The least squares line is $\hat{y} = 3.10 + 2.76\sqrt{x}$
(Using rounded values $3.097869 \approx 3.10$ and $2.7633138 \approx 2.76$).
That is, Estimated Time Needed $= 3.10 + 2.76\sqrt{\text{Number of Items}}$.

11.9 a. A scatterplot of the data is given here:

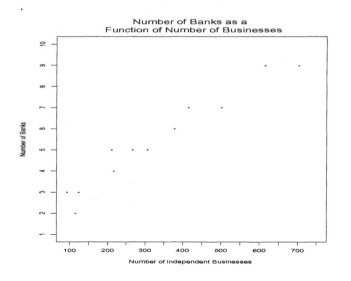

Number of Banks as a
Function of Number of Businesses

From a scatterplot of data, a straight-line relationship between y and x seems reasonable.

b. The intercept is 1.766846 (1.77) and the slope is 0.0111049 (0.0111).

c. An interpretation of the estimated slope is as follows. For an increase of 1 business in a zip code area, there is an increase of 0.0111 in the average number of banks in the zip code.

d. From the output, the sample residual standard deviation is given by "Root MSE", 0.5583.

11.11 a. The **LOWESS** smooth is basically a straight line, with a slight bend at the larger values of Income. Thus, it would appear that the relation is basically linear.

b. The points with high leverage are those that are far from the average value of the x-values along the x-axis. In this case there are two points with very high income values. They are the last two values in the data listing. The plot shows that the point with Income=65.0 and Price=110.0 is a considerable distance from the **LOWESS** line. This point therefore has high influence. The other high-leverage point Income=70.0 and Price=185.0 falls reasonably close to the **LOWESS** line. This point would not have such a large influence on the fitted line.

11.13 The estimated slope changed considerably from 1.80 to 2.46. This resulted from excluding the data value having high-influence. Such a data point twists the fitted line towards itself and hence can greatly distort the value of the estimated slope.

11.15 p-value < 0.0001

11.17 Minitab Output is given here:

```
Regression Analysis: LogRecovery versus Time


The regression equation is
LogRecovery = 1.67 - 0.0159 Time

Predictor        Coef      SE Coef         T        P
Constant      1.67243      0.05837     28.65    0.000
Time         -0.015914    0.001651     -9.64    0.000


S = 0.1114      R-Sq = 89.4%     R-Sq(adj) = 88.5%

Analysis of Variance

Source           DF          SS          MS        F        P
Regression        1      1.1523      1.1523    92.93    0.000
Residual Error   11      0.1364      0.0124
Total            12      1.2887
```

    a. $\hat{y} = 1.67 - 0.0159x$

    b. $s_\epsilon = 0.1114$

    c. $SE(\hat{\beta}_o) = 0.05837 \quad SE(\hat{\beta}_1) = 0.001651$

11.19   a. Yes, the data values fall approximately along a straight-line.

       b. $\hat{y} = 12.51 + 35.83x$

11.21   a. Scatterplot of the data is given here:

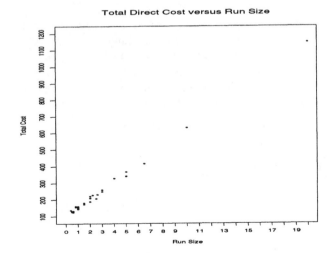

Total Direct Cost versus Run Size

An examination of the scatterplot reveals that a straight-line equation between total cost and run size may be appropriate. There is a single extreme point in the data set but no evidence of a violation of the constant variance requirement.

b. $\hat{y} = 99.777 + 5.1918x$

   The residual standard deviation is $s = \sqrt{148.999} = 12.2065$

c. A 95% C.I. for the slope is given by $\hat{\beta}_1 \pm t_{0.025,28} SE(\hat{\beta}_1) \Rightarrow 5.1918 \pm (2.048)(0.586455) \Rightarrow$
   $(5.072, 5.312)$

11.23 a. $F = 7837.26$ with p-value $= 0.0000$.

   b. The F-test and two-sided t-test yield the same conclusion in this situation. In fact, for this type of hypotheses, $F = t^2$.

11.25 95% prediction interval for log biological recovery percentage at x=30 is given by

$$\hat{y} \pm t_{.025,11} s_\epsilon \sqrt{1 + \tfrac{1}{n} + \tfrac{(30-\bar{x})^2}{S_{xx}}} \Rightarrow$$

$1.195 \pm (2.201)(0.1114)\sqrt{1 + \tfrac{1}{13} + \tfrac{(30-30)^2}{4550}} \Rightarrow 1.195 \pm 0.0.254 \Rightarrow (0.941, 1.449)$

The prediction interval is somewhat wider than the confidence interval on the mean.

11.27 a. 95% Confidence Intervals for $E(y)$ at selected values for x:
   $x = 4 \Rightarrow (2.6679, 4.3987)$
   $x = 5 \Rightarrow (4.2835, 5.4165)$
   $x = 6 \Rightarrow (5.6001, 6.7332)$
   $x = 7 \Rightarrow (6.6179, 8.3487)$

   b. 95% Prediction Intervals for y at selected values for x :
   $x = 4 \Rightarrow (1.5437, 5.5229)$
   $x = 5 \Rightarrow (2.9710, 6.7290)$
   $x = 6 \Rightarrow (4.2877, 8.0456)$
   $x = 7 \Rightarrow (5.4937, 9.4729)$

   c. The confidence intervals in part (a) are interpreted as "We are 95% confident that the average weight loss over many samples of the compound when exposed for 4 hours will be between 2.67 and 4.40 pounds." Similar statements for other hours of exposure.

   The prediction intervals in part (b) are interpreted as "We are 95% confident that the weight loss of a single sample of the compound when exposed for 4 hours will be between 1.54 and 5.52 pounds." Similar statements for other hours of exposure.

11.29 No, because $x = 2.0$ is close to the mean of all x-values used in determining the least squares line.

11.31 a. Scatterplot of the data is given here:

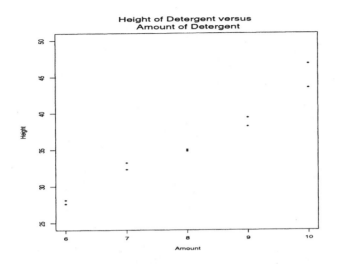

b. The SAS output is given here:

```
Model: MODEL1
Dependent Variable: Y
```

Analysis of Variance

Source	DF	Sum of Squares	Mean Square	F Value	Prob>F
Model	1	330.48450	330.48450	169.213	0.0001
Error	8	15.62450	1.95306		
C Total	9	346.10900			

Root MSE	1.39752	R-square	0.9549	
Dep Mean	35.89000	Adj R-sq	0.9492	
C.V.	3.89390			

Parameter Estimates

Variable	DF	Parameter Estimate	Standard Error	T for H0: Parameter=0	Prob > |T|
INTERCEP	1	3.370000	2.53872138	1.327	0.2210
X	1	4.065000	0.31249500	13.008	0.0001

$\hat{y} = 3.37 + 4.065x$

c. The residual plot is given here:

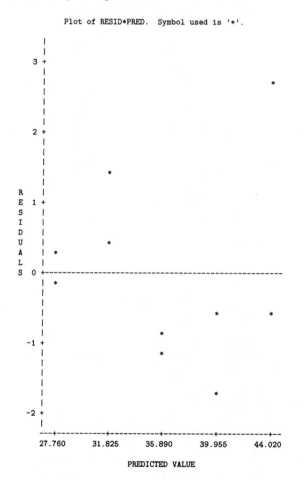

Plot of RESID*PRED. Symbol used is '*'.

The residual plot indicates that higher order terms in x may be needed in the model.

11.33 $\hat{y} = 1.47 + 0.797x$

$H_o : \beta_1 \leq 0$ versus $H_a : \beta_1 > 0$

$t = 7.53 \Rightarrow$ p-value $= Pr(t_8 \geq 7.53) < 0.0005 \Rightarrow$

There is sufficient evidence in the data that the slope is positive.

11.35   a. $\hat{y} = 0.11277 + 0.11847x \approx 0.113 + 0.118x$

Dependent variable is Transformed CUMVOL and Independent variable is Log(Dose).

b. $\hat{x} = (y - 0.11277)/0.11847$

$s_\epsilon = 0.04597, S_{xx} = 4.9321, \bar{x} = 2.40, t_{.005,8} = 2.819 \Rightarrow$

$c^2 = \frac{(2.819)^2(.04597)^2}{(.11847)^2(4.9321)} = 0.2426$

$d = \frac{(2.819)(.04597)}{.11847} \sqrt{\frac{24+1}{24}(1 - .242467) + \frac{(\hat{x}-2.40)^2}{4.9321}}$

70

$$\hat{x}_L = 2.40 + \frac{1}{(1-.2426)}(\hat{x} - 2.40 - d)$$
$$\hat{x}_U = 2.40 + \frac{1}{(1-.2426)}(\hat{x} - 2.40 + d) \Rightarrow$$

y	TRANS(y)	$\widehat{LOG(x)}$	$\hat{x}$	d	$\widehat{LOG(x)}_L$	$\widehat{LOG(x)}_U$	$\hat{x}_L$	$\hat{x}_U$
10	.322	1.764	5.84	.242289	1.24038	1.88018	3.46	6.55
14	.383	2.285	9.83	.198634	1.98616	2.51068	7.29	12.31
19	.451	2.855	17.38	.221359	2.70876	3.29328	15.01	26.93

11.37 The output yields R-square=0.9452. The estimated slope of the regression line is 0.0111049 which is positive, indicating an increasing relation between Branches and Business. Thus, the correlation is the positive square root of 0.9452, i.e., $r = 0.9722$.

11.39 a. The correlation is given as 0.956. This value indicates a strong positive trend between intensity of the advertising and the awareness percentage. That is, as intensity of the advertising increases we would generally observe an increase in awareness percentage.

b. Scatterplot of the data is given here:

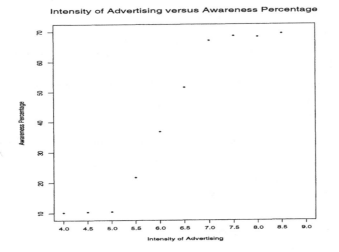

The relation between intensity and awareness does not appear to follow a straight line. It is generally an increasing relation with a threshold value of approximately x=5 before awareness increases beyond a value of 10%. A leveling off of the sharp increase in awareness occurs at an intensity of x=7.

11.41 For the points with the smaller symbols, there is a general upward trend in SALARY as EXPER increases in value. The point denoted with a large box, SALARY=37.9 and EXPER 14 years, clearly does not fall within the general pattern of the other points. This point is an outlier; it has high leverage because the experience for this graduate is considerably larger than the average, and the point has high influence because the starting salary value is much smaller than all the other points having experience value close to 14 years.

71

11.43   a. $\hat{y} = 5.890659 + 0.0148652x$

   b. $H_o : \beta_1 \leq 0$ versus $H_a : \beta_1 > 0$, from the output $t = 5.812$.

   The p-value on the output is for a two-sided test. Thus, p-value $= Pr(t_5 \geq 5.812) = 0.00106$ which indicates that there is significant evidence that the slope is greater than 0.

11.45 The regression line using ln(x) appears to provide the better fit. The scatterplots indicate that the line using ln(x) more closely matches the data points. Also, the residual standard deviation using ln(x) is smaller than the standard deviation using x (2.0135 vs 2.3801).

11.47   a. The prediction equation is $\hat{y} = 140.074 + 0.61896x$.

   b. The coefficient of determination is $R^2 = 0.9420$ which implies that 94.20% of the variation in fuel usage is accounted for by its linear relationship with flight miles. Because the estimated slope is positive, the correlation coefficient is the positive square root of R^2, i.e., $r = \sqrt{0.9402} = 0.97$.

   c. The only point in testing $H_o : \beta_1 \leq 0$ versus $H_a : \beta_1 > 0$ would be in the situation where the flights were of essentially the same length and there is an attempt to determine if there are other important factors that may affect fuel usage. Otherwise, it would be obvious that longer flights would be associated with greater fuel usage.

11.49   a. The group felt that towns with small populations should be associated with large amounts per person, whereas towns with larger populations would have small amounts per person. Thus, the group claimed that as townsize increased the amount spent per person decreased. If the data supported their claim, then the slope would be negative.

   b. The estimated slope was $\hat{\beta}_1 = 0.0005324$ which is positive. Thus, there claim is not supported by the data.

11.51   a. The point is a very high-influence outlier which has distorted the slope considerably.

   b. The regression line with the one point eliminated has a negative slope, $\hat{\beta}_1 = -0.0015766$. This confirms the opinion of the group, which had argued that the smallest towns would have the highest per capita expenditures with decreasing expenditures as the size of the towns increased.

11.53   a. From the scatterplot, it would appear that a straight line model relating Homogenate to Pellet would provide an adequate fit.

   b. No, the plot does not indicate any outliers and the variance appears to be constant.

   c. We could use the calibration techniques and predict Pellet response using the observe Homogenate reading.

   d. Use the calibration prediction interval.

Chapter 12: Multiple Regression

12.1   a. A scatterplot of the data is given here:

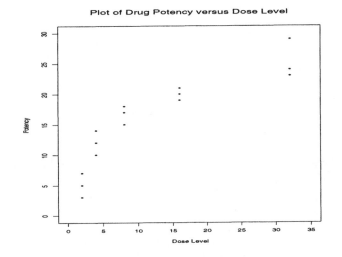

Plot of Drug Potency versus Dose Level

b. $\hat{y} = 8.667 + 0.575x$

c. From the scatterplot, there appears to be curvature in the relation between Potency and Dose Level. A quadratic or cubic model may provide an improved fit to the data.

d. The quadratic model provides the better fit. The quadratic model has a much lower MS(Error), its R^2 value is 11% larger, the quadratic term has a p-value of 0.0062 which indicates that this term is significantly different from 0, however, the residual plot still has a distinct curvature as was found in the residual plot for the linear model.

12.3   a. A scatterplot of the data is given here:

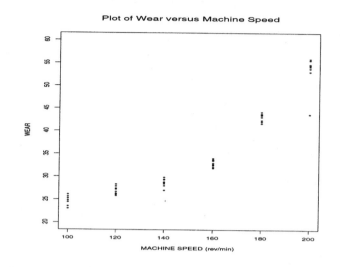

Plot of Wear versus Machine Speed

b. Since there is curvature in the relation, a quadratic model would be a possibility.

c. The quadratic model gives a better fit. For the quadratic model the residual plots displays a slight pattern in the residuals but not as distinct as found in the residual plot from the linear model. The R^2 values from the quadratic and cubic models are nearly identical but are about 10% higher than the value from the linear model. The cubic term has p-value=0.1794 which would indicate that the cubic term is not making a significant contribution to the fit of the model above just using a model having linear and quadratic terms.

d. There is one data value (Machine Speed=200, Wear=43.7) which is definitely an outlier. Considering the variability in the Wear values at each Machine Speed, it is possible that there is an error in recording the Wear value and in fact it should be 53.7. A check with the lab personnel would need to be done.

12.5 a. No, the two independent variables, Air Miles and Population, do not appear to be severly collinear, based on the correlation (-0.1502) and the scatterplot.

b. There are two potential leverage points in the Air Miles direction (around 300 and 350 miles). In addition, there is one possible leverage point in the population direction; this point has a value above 200.

12.7 a. For reduced model: $R^2 = 0.2049$

b. For complete model: $R^2 = 0.7973$

c. With INCOME as the only independent variable, there is a dramatic decrease in R^2 to a relatively small value. Thus, we can conclude that INCOME does not provide an adequate fit of the model to the data.

12.9 a. $R^2 = 0.979566 \Rightarrow 97.96\%$ of the variation in Rating Score is accounted for the model containing the three independent variables.

b. $F = \frac{0.979566/3}{(1-0.979566)/496} = 7925.76$ This value is slightly smaller than the value on the output due to rounding-off error.

c. The p-value associated with such a large F-value would be much less than 0.0001 and hence there is highly significant evidence that the model containing the three independent variables provides predictive value for the Rating Score.

12.11   a. $\hat{y} = 50.0195 + 6.64357x_1 + 7.3145x_2 - 1.23143x_1^2 - 0.7724x_1x_2 - 1.1755x_2^2$

b. $\hat{y} = 70.31 - 2.676x_1 - 0.8802x_2$

c. For the complete model: $R^2 = 86.24\%$
   For the reduced model: $R^2 = 58.85\%$

d. In the complete model, we want to test
   $H_o : \beta_3 = \beta_4 = \beta_5 = 0$ versus H_a : at least one of $\beta_3, \beta_4, \beta_5 \neq 0$.
   The F-statistic has the form:
   $$F = \frac{[SSReg.,Complete - SSReg.,Reduced]/(k-g)}{SSResidual,Complete/[n-(k+1)]} = \frac{[448.193 - 305.808]/(5-2)}{71.489/[20-6]} = 9.29$$
   with $df = 3, 14 \Rightarrow$ p-value $= Pr(F_{3,14} \geq 9.29) = 0.0012 \Rightarrow$
   Reject H_o. There is substantial evidence to conclude that at least one of $\beta_3, \beta_4, \beta_5 \neq 0$. Based on the F-test, omitting the second order terms from the model has substantially changed the fit of the model. Dropping one or more of these independent variables from the model will result in a decrease in the predictive value of the model.

12.13   a. The Minitab output for fitting the complete and reduced models is given here:

```
Regression Analysis: y versus x1, x2, x3, x1*x2, x1*x3

The regression equation is
y = 7.31 + 3.30 x1 - 2.15 x2 - 4.35 x3 - 1.50 x1*x2 - 2.28 x1*x3

Predictor       Coef      SE Coef        T         P
Constant       7.3072      0.2103      34.75     0.000
x1             3.3038      0.2186      15.11     0.000
x2            -2.1548      0.2974      -7.25     0.000
x3            -4.3486      0.2974     -14.62     0.000
x1*x2         -1.5004      0.3092      -4.85     0.003
x1*x3         -2.2795      0.3092      -7.37     0.000

S = 0.3389     R-Sq = 98.8%     R-Sq(adj) = 97.7%

Analysis of Variance

Source           DF        SS         MS        F        P
Regression        5     55.293     11.059    96.30    0.000
Residual Error    6      0.689      0.115
Total            11     55.982
```

```
Regression Analysis: y versus x1, x2, x3

The regression equation is
y = 6.59 + 2.04 x1 - 1.30 x2 - 3.05 x3

Predictor        Coef     SE Coef        T       P
Constant       6.5894      0.5131    12.84   0.000
x1             2.0438      0.3519     5.81   0.000
x2            -1.3000      0.6679    -1.95   0.087
x3            -3.0500      0.6679    -4.57   0.002

S = 0.9446      R-Sq = 87.2%      R-Sq(adj) = 82.5%

Analysis of Variance

Source           DF        SS         MS       F       P
Regression        3    48.844     16.281   18.25   0.001
Residual Error    8     7.138      0.892
Total            11    55.982
```

In the complete model: $y = \beta_o + \beta_1 x_1 + \beta_2 x_2 + \beta_3 x_3 + \beta_4 x_1 x_2 + \beta_5 x_1 x_3 + \epsilon$, the test of equal slopes is a test of the hypotheses:

$H_o : \beta_4 = 0, \beta_5 = 0$ versus $H_o : \beta_4 \neq 0$ and/or $\beta_5 \neq 0$

Under H_o, the reduced model becomes $y = \beta_o + \beta_1 x_1 + \beta_2 x_2 + \beta_3 x_3 + \epsilon$

$F = \frac{(55.293 - 48.844)/(5-3)}{0.689/6} = 28.08 \Rightarrow$ p-value $= Pr(F_{2,6} \geq 28.08) = 0.0009$

b. Reject H_o and conclude there is significant evidence that the slopes of the three regression lines (one for each Drug Product) are different.

c. In the complete model, a test of equal intercepts is a test of the hypotheses:

$H_o : \beta_2 = 0, \beta_3 = 0$ versus $H_o : \beta_2 \neq 0$ and/or $\beta_3 \neq 0$

Under H_o, reduced model becomes $y = \beta_o + \beta_1 x_1 + \beta_4 x_1 x_2 + \beta_5 x_1 x_3 + \epsilon$

Obtain the SS's from the reduced model and then conduct the F-test as was done in part a.

12.15   a. For testing $H_o : \beta_1 = 0$ versus $H_a : \beta_1 \neq 0$, the p-value for the output is p-value < 0.0001. Thus, we can reject H_o and conclude there is significant evidence that the probability of Tumor Development is related to Amount of Additive .

b. From the output, $\hat{p}(100) = 0.827$ with 95% C.I. (0.669, 0.919).

12.17 The output for fitting a regression model having TravTime regressed on Miles, TravDir, DirEffct is given here.

```
Regression Analysis: TravTime versus Miles, TravDir, DirEffct

The regression equation is
TravTime = 0.645 + 0.00191 Miles - 0.0186 TravDir -0.000079 DirEffct

Predictor         Coef      SE Coef         T        P
Constant       0.64494      0.02635     24.47    0.000
Miles       0.00190703   0.00002034     93.74    0.000
TravDir       -0.01859      0.02065     -0.90    0.370
DirEffct   -0.00007890   0.00001265     -6.24    0.000

S = 0.1609     R-Sq = 98.9%     R-Sq(adj) = 98.9%

Analysis of Variance

Source            DF          SS         MS       F        P
Regression         3     231.100     77.033  2975.60    0.000
Residual Error    96       2.485      0.026
Total             99     233.585

Source         DF      Seq SS
Miles           1     226.643
TravDir         1       3.449
DirEffct        1       1.008

Unusual Observations
Obs     Miles   TravTime        Fit     SE Fit    Residual    St Resid
 11      4501     9.5833     9.9759     0.1088     -0.3926      -3.31RX
 13      1746     3.9667     4.2873     0.0336     -0.3206      -2.04R
 39      1561     4.0833     3.7636     0.0232      0.3198       2.01R
 54      3784     7.0000     7.2269     0.0872     -0.2269      -1.68 X
 64      2477     5.3333     4.9406     0.0479      0.3927       2.56R
 71      1395     4.0833     3.4339     0.0215      0.6494       4.07R

R denotes an observation with a large standardized residual
X denotes an observation whose X value gives it large influence.

Predicted Values for New Observations

New Obs     Fit     SE Fit        95.0% CI            95.0% PI
1        1.7145     0.0356   ( 1.6438,  1.7852)   ( 1.3874,  2.0416)

Values of Predictors for New Observations

New Obs    Miles   TravDir   DirEffct
1            500     -2.00      -1000
```

When DirEffct is included in a model with TravDir, the variable TravDir no longer provides a significant contribution to the fit of the model (p-value=0.37). Therefore, with DirEffct in the model, it is not necessary to also include the variable TravDir. This conclusion is further confirmed by comparing the models with Miles and just DirEffct to the model with Miles, DirEffct, and TravDir. The model without TravDir has the same value for R^2_{adj} as the model with TravDir, but a much larger value for the F-test of model significance (4471.71 vs 2975.60). From the output, the predicted travel time would be 1.7145 hours with a 95% prediction interval of (1.3874, 2.0416).

12.19 The results of the various t-tests are given here:

H_o	H_a	T.S. t	p-value	Conclusion
$\beta_o = 0$	$\beta_o \neq 0$	$t = -0.02$	0.982	Fail to Reject H_o
$\beta_1 = 0$	$\beta_1 \neq 0$	$t = 2.05$	0.079	Fail to Reject H_o
$\beta_2 = 0$	$\beta_2 \neq 0$	$t = 0.68$	0.520	Fail to Reject H_o
$\beta_3 = 0$	$\beta_3 \neq 0$	$t = 1.40$	0.203	Fail to Reject H_o
$\beta_4 = 0$	$\beta_4 \neq 0$	$t = -1.71$	0.131	Fail to Reject H_o

None of the four independent variables appears to have predictive value given the remaining three variables have already been included in the model.

12.21   a. The regression model is

$\hat{y} = -16.8198 + 1.47019x_1 + .994778x_2 - .0240071x_3 - .01031x_4 - .000249574x_5$

$s_\epsilon = 3.39011$

     b. Test $H_o : \beta_3 = 0$ versus $H_a : \beta_3 \neq 0$. From output, $t = -1.01$ with p-value=0.3243. Thus, there is not substantial evidence that the variable $x_3 = x_1x_2$ adds predictive value to a model which contains the other four independent variables.

12.23   a. $\hat{y} = 102.708 - .833$ PROTEIN $- 4.000$ ANTIBIO $- 1.375$ SUPPLEM

     b. $s_\epsilon = 1.70956$

     c. $R^2 = 90.07\%$

     d. There is no collinearity problem in the data set. The correlations between the pairs of independent variables is 0 for each pair and the VIF values are all equal to 1.0. This total lack of collinearity is due to the fact that the independent variables are perfectly balanced. Each combination of PROTEIN and ANTIBIO values appear exactly three times in the data set. Each combination of PROTEIN and SUPPLEM occur twice, etc.

12.25   a. $\hat{y} = 89.8333 - 0.83333$ PROTEIN

     b. $R^2 = 0.5057$

     c. In the complete model, we want to test

$H_o : \beta_2 = \beta_3 = 0$ versus $H_a :$ at least one of $\beta_2, \beta_3 \neq 0$.

The F-statistic has the form:

$F = \frac{[371.083 - 208.333]/(3-1)}{40.9166/[18-4]} = 27.84$

with $df = 2, 14 \Rightarrow$ p-value $= Pr(F_{2,14} \geq 27.84) < 0.0001 \Rightarrow$ Reject H_o.

There is substantial evidence to conclude that at least one of $\beta_2, \beta_3 \neq 0$. Based on the F-test, omitting x_2 and/or x_3 from the model would substantially changed the fit of the model. Dropping ANTIBIO and/or SUPPLEM from the model may result in a large decrease in the predictive value of the model.

12.27   a. The box plots of the data are given here:

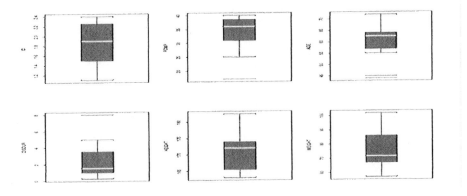

b. The scatterplots of the data are given here:

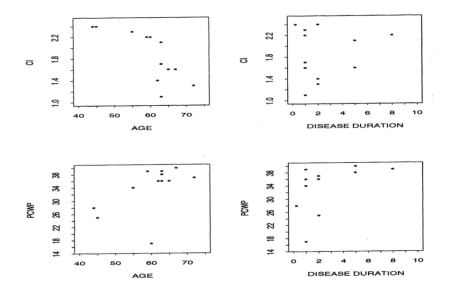

There appears to be somewhat of a negative correlation between Age and CI but a positive correlation between Age and PCWP. The relation between Disease Duration and CI is very weak but slightly positive. While the correlation between disease duration and PCWP is somewhat stronger.